2.50

FOREIGN & COMMONW...

United States Interests and Policies in Africa

Also by Karl P. Magyar

CHALLENGE AND RESPONSE: Anticipating US Military Security Concerns (*editor*)

CONFLICT, CULTURE AND HISTORY: Regional Dimensions (*co-editor*)

GLOBAL SECURITY CONCERNS: Anticipating the Twenty-First Century (*editor*)

THE INDIAN SOUTH AFRICANS: A Contemporary Profile (*co-editor*)

* PEACEKEEPING IN AFRICA (*editor with Earl Conteh-Morgan*)

PROLONGED WARS: The Post-Nuclear Challenge (*co-editor*)

* *From the same publishers*

United States Interests and Policies in Africa

Transition to a New Era

Edited by

Karl P. Magyar
Professor of National Security Affairs
United States Air Command and Staff College
Alabama

First published in Great Britain 2000 by
MACMILLAN PRESS LTD
Houndmills, Basingstoke, Hampshire RG21 6XS and London
Companies and representatives throughout the world

A catalogue record for this book is available from the British Library.

ISBN 0–333–69447–3

First published in the United States of America 2000 by
ST. MARTIN'S PRESS, INC.,
Scholarly and Reference Division,
175 Fifth Avenue, New York, N.Y. 10010

ISBN 0–312–22388–9

Library of Congress Cataloging-in-Publication Data
United States interests and policies in Africa : transition to a new
era / edited by Karl P. Magyar.
p. cm.
Includes bibliographical references and index.
ISBN 0–312–22388–9
1. United States—Foreign relations—Africa.   2. Africa—Foreign
relations—United States.   I. Magyar, K. P. (Karl P.)
JZ1480.U554   1999
327.7306'09'049—dc21   99–15387
                                                                    CIP

This book is printed on paper suitable for recycling and made from fully managed and sustained
forest sources.

10   9   8   7   6   5   4   3   2   1
09   08   07   06   05   04   03   02   01   00

Printed and bound in Great Britain by
Antony Rowe Ltd, Chippenham, Wiltshire

Contents

Preface

As each historical era collapses and before the character of a new one is identified, global relations among states, nations, cultures, and civilizations enter a period of intense transition. In the absence of a grand architect who would design and erect such an era single handedly, a vast global army of history scholars and analysts are yet again seeking to understand the various highly intricate dimensions of the defunct era. Recently our combined analytic efforts evidently failed once again to grasp the major underlying currents of change responsible for the demise of the Cold War era. The speed with which the Soviet Union collapsed ought to be of serious concern to professional analysts. Historians will examine the lacunae for any evidence that should have served as beacons for major, revolutionary changes that occurred. Why was this not perceived? Other analytic specialists will scramble to grasp the nature of the newly emerging system. Will it be the great Marxist dialectical synthesis? A religious 'Kingdom of God on earth'? A global *Pax Americana*? The end of history? I have confidence that history will yet again demonstrate its independent power by dictating events to which the humble, but wise, will respond, and not aspire to engineer a revolutionary, handcrafted, human utopia – in pursuit of universal justice.

There is irony in this situation. The Marxists had envisaged the proletarian revolution as the medium through which the workers' vanguard begins the task of cleansing the capitalist sources of men's historical abuses. For Marxists, man has the power to make the revolution that in turn allows the socialist man to design his own eschatological destiny. The gods were not necessary for this and, indeed, they were also ranked among the counter-revolutionaries because the gods were perceived as allies of the entrenched capitalists who enslaved the alienated workers whose only existential purpose was to serve the degenerate and acquisitive owners of the means of production and their political running dogs. Hence the positive attitude towards the proletarian revolution. The irony emerges in the form of the western penchant for social revolutions, and in the United States for the highly regarded attitudes evinced in our management culture towards problem solvers, can-do producers, and creators of huge political and economic enterprises – all skills that are attributed rather serenely to management education. However, when our managers fail, public relations campaigns are invoked to undo cosmetically at least the visible damage. The damage only is addressed; rarely the cause. For the Marxists, the waning capitalists could be blamed for failures, but certainly not those in the vanguard of the proletarian revolution. In this sense, the paragon of exploitative capitalists could be considered to be methodological allies of Marxist revolutionaries in that both validated man's (society's) ability to chart its own course and destiny by the implementation of pseudo-scientific analytic devices, knowable to man. Marxists and modern capitalist managers were true ideological optimists. And both have been, and remain perplexed with Africa.

This book concerns changing United States relations with Africa. As with many other academic topics undergoing examination for change in the present transitional period, the perceived gravity of change, if any, remains largely a judgmental exercise in subjectivity due to the inherently social nature of the inquiry. Weighing the positive against the negative factors to be judged might, presumably, be preceded by a joint methodological exercise among all participants involved in this analytic effort, yet it would not necessarily accrue objectivity to the reader. Hence my choice was to allow for individual interpretations of each of the five regions of Africa that we examined. Even the identification and inclusion of states within a region posed dilemmas no doubt familiar to all Africa analysts. Does Sudan properly belong to North or East Africa? Should Mauritania be classified as a part of North or West Africa? Where do Rwanda and Burundi belong? And, does Southern Africa have 10, 11, or 13 members – as postulated by various analyses? As Editor, I provided loose guidance but I did not dictate compliance. Indeed, in this example alone, changes taking place in Africa in the post-Cold War era might soon influence arguments for new sub-regional alignments or analytic paradigms. Should Sudan undergo political and territorial fragmentation, the placement of the segments might facilitate the decision concerning their proper regional identities.

Ours' are assessments based on individual evaluations, hence as Editor I did not require a common view concerning the degree of change, nor the nature of it. The participants in this analytic project are well-established Africa specialists and therefore their individual attitude towards *change* will emerge in their presentations. We had the good fortune of convening at several professional conferences but, again, we adopted no common attitude towards the main subject of our inquiry. An imposed common framework might have encouraged great distortions in the estimate of the relevance of the five regions to the United States. We accept that we are supplying early, broad, chapter-length agendas that present the minimal relevant factors of each region to be considered perhaps for longer elaborations in the future. But identifying such major factors cannot be an objective exercise.

Those who are professionally concerned with the study of Africa's affairs are well familiar with the role of perceptions which has spawned the dichotomy of 'Afro-optimists' and 'Afro-pessimists'. We expect the reader will make judgments regarding our individual presentations accordingly. Anyone examining African affairs cannot escape the 'good news, bad news' dilemma. An example of this phenomenon was illustrated by President Clinton's 1998 12-day tour through Africa during which he enthused about a present-day Africa undergoing a 'renaissance', replete with high economic growth rates, stabilization, and democratization. However, shortly after that unprecedented official visit, US Commerce Secretary William Daley soberly noted that only one percent of America's exports go to Africa, and that in the last 40 years, Africa's share of world trade fell by half. Not long after Clinton's brief stop in Rwanda, mass killings started again; Ethiopia, whose Prime Minister attended the regional forum in Uganda that met with Clinton and signed measures to work towards peace, engaged in bombing several disputed positions in Eritrea a few months after that meeting; Senegal had recently been identified as resorting to torture and killings in the attempt to keep control over its secessionist province of Casamance; and most of the heads of state with which Clinton met, had been in power for a very long time, or had come to power by military take over. Ironically, it was in the notorious apartheid South Africa, in which Nelson Mandela came to power legitimately. Such are the evaluative dilemmas faced routinely by analysts of Africa's affairs. Therein lies the caution leveled at the reader not familiar with that otherwise fabulous continent. Grand judgments concerning Africa had best be avoided.

As Drs Winkates, Copson, and Magyar are employed by US Government agencies, we emphasize that the views presented in this volume are the private views of the authors and do not necessarily represent the official positions of our specific, or any, US Government agencies. The Editor is pleased to express his gratitude to Ms Gwen Story, Supervising Writer-Editor, and her staff, Ms Yuna Braswell, Ms Becky Bryant, and Ms Julie Hall, Editorial Assistants at Air University's Air Command and Staff College. I will forever thank Ms Annabelle Buckley, Senior Commissioning Editor at Macmillan, for her efforts on our behalf and also for her admirable patience.

Karl P. Magyar

Notes on the Contributors

Earl Conteh-Morgan, born in Sierra Leone, holds a PhD from Northwestern University. He is Associate Professor of International Studies at the University of South Florida. Dr Conteh-Morgan has published extensively on North–South relations, conflict processes, and value changes in developing countries. Author of several books and numerous articles on international economics, democratization, and political transition, he was a 1995 Senior Research Fellow at the Norwegian Nobel Institute.

Raymond W. Copson, prior to his arrival at the Congressional Research Service in 1978, lectured at universities in Kenya and Nigeria. He is currently Head of the Europe/Middle East/Africa section of the Foreign Affairs and National Defense Division at CRS. Dr Copson holds a PhD in International Studies from the Johns Hopkins University. His current research interests are in the area of Africa's conflicts and international relations. His recent book *Africa's Wars and Prospects for Peace* (1994) was supported by the United States Institute of Peace and the National Defense University.

Karl P. Magyar, the editor, is Professor of National Security Affairs at the US Air Force Air Command and Staff College. He holds a PhD in International Relations from the Johns Hopkins University. Dr Magyar has held academic positions in the United States, with the US Air Force in Germany, Italy, and Great Britain, as visiting professor in Japan, and taught five years in South Africa. In addition, he has served as International Trade Specialist at the US Department of Commerce in Washington and Nigeria.

James E. Winkates is Research Professor of International Affairs and Senior Curriculum Advisor at the US Air Force Air War College. He holds a PhD in International Affairs from the University of Virginia. He has visited East Africa on several occasions, most recently in 1995 when he served as academic advisor to the first Air War College delegation to the region. Dr Winkates's academic specialties include US foreign and defense policies, terrorism, and an Africa area focus.

Mohamad Z. Yakan, having earned a PhD in political science at the University of Michigan, served on the Faculties of the Lebanese American University, Wayne State University, and the University of San Diego. He is currently Associate Professor of International Relations at United States International University in San Diego. Dr Yakan's active teaching and research background includes Islamic Thought and Jurisprudence, and a developed focus on Middle Eastern and African affairs.

List of Abbreviations

ACRI	African Crisis Response Initiative
ADFL	Alliance of Democratic Forces for the Liberation of Congo–Zaire
AMU	Arab Maghreb Union
AUC	American University of Cairo
CAR	Central African Republic
CENTCOM	US Central Command
CENTO	Central Treaty Organization
COMECON	Council for Mutual Economic Assistance
ECOMOG	ECOWAS Ceasefire Monitoring Group
ESF	Economic Support Fund
FIS	Islamic Salvation Front
FLN	Algerian Front de Liberation National
FRELIMO	*Frente de Liberatacao de Moçambique*
IFC	International Finance Corporation
IMETP	International Military Education and Training Program
IMF	International Monetary Fund
INSS	Institute for Strategic Studies
MAP	Military Assistance Program
MOOTW	Military Operations Other Than War
MPLA	Movement for the Liberation of Angola
NATO	North Atlantic Treaty Organization
NGO	Non-Governmental Organization
NSC	National Security Council
NSR	National Security Review
NSSM	National Security Study Memorandum
OAU	Organization of African Unity
OECD	Organization for Economic Cooperation and Development
OPEC	Organization of Petroleum Exporting Countries
PFLP	Popular Front for the Liberation of Palestine
PLO	Palestine Liberation Organization
PRC	People's Republic of China
RPF	Rwanda Patriotic Front
SADC	Southern African Development Community
SAP	IMF Structural Adjustment Program
SWAPO	South West Africa People's Organization

UN	United Nations
UNAMIR	United Nations Assistance Mission in Rwanda
UNISOM	United Nations Operations in Somalia
UNITA	*Unaio Nacional para a Independencia Total de Angola*
USAID	US Agency for International Development

Introduction: Africa's Transitional Role in America's Post-Cold War Era Diplomacy

Karl P. Magyar

Historical Perspectives

America's interests in Africa date back to 1786, at which time Morocco officially recognized the new United States of America. Within six years, the United States had established a consulate, and soon thereafter, other legations were opened along the North African coast. The United States also engaged militarily the Barbary pirates. During the following century, America's diplomatic interests in Africa concerned mostly the Portuguese territories, the Congo, and Liberia – which had been established as a bit of an American cultural outpost by freed American slaves. At the Berlin Conference in 1884–5 at which Africa was carved up among the European colonial powers, the United States participated mostly in an observer capacity, and expressed some humanitarian and commercial concerns. Except for these modest activities, and the concern expressed about Ethiopia's fate in the face of Italy's military incursion prior to World War II, the United States had traditionally viewed Africa mostly out of anthropological curiosity, expressed more often in the pages of *National Geographic* magazine than in government documents regarding America's diplomatic interests in the world.

This isolation from Africa's affairs changed rapidly when the flood of newly available, independent African states became the object of the expanding superpower competition in the Cold War. Aggressive Soviet advances on European territory had been stalemated, and the iron curtain would be kept in place, but not rolled back by force. The Soviets attempted to build their politico-strategic alliance in order to expand their global network of support in the relentless conflict with global capitalism. The United States in turn recognized the limited value of adding these numerous new states to its camp, but *denying* them to the other side was sufficient reason for establishing active American interests in Africa. Thus was born Africa's peripheral role in the affairs of the Cold War world. Certainly Western Europe retained its long-established relationship with the African continent but this concerned colonial history, culture, Christianity, and commerce. In matters of security strategy, the Soviets and Americans competed for advantage but with whatever African security interests which may have existed playing a distinctly lesser role to that of the focal core Cold War region: continental Europe.

During the last four decades, Africa's lesser level in America's global diplomatic concerns has been reflected in generally inconclusive evaluations by analysts. Thus William J. Foltz lists Africa's five roles in great power strategic calculations:

- Physical obstacle to other points.
- Defensive bastion to protect sea lanes.
- Launching pad for attacks on other territory.
- Source of military supplies.
- A surrogate battle terrain.[1]

Notable in his analysis is the stress on Africa's physical geography, with no item referring to a humanitarian dimension. The physical continent features more than its human inhabitants do.

David A. Dickson portrays the Eisenhower administration's primacy of military-security issues over economic and political development of the fragile states in Africa.[2] However, Melvin Gurtov offers a contrasting perspective, 'American response to Africa was one of studied neglect'.[3] Africa's value to the United States was only economic, mostly oil and strategic minerals – which should be denied to the communist nations. Then Vice-President Nixon noted at least Africa's investment potential. He also counseled that the United States should back Africa's long-standing ties to Europe. These early contrasting estimates of Africa's interest to the United States structured the debate that was clearly won by those with a global view of the Cold War. Accordingly, the United States developed three air bases in North Africa and the Azores and gained access to several port facilities in Africa.

These military advantages were not reversed by the Kennedy administration, but as Gurtov notes, Chester Bowles, Kennedy's Under-Secretary of State, called for a 'revision of tactics, not objectives' – which recognized Africa's calls for self-determination, rather than backing Europe's assumed perpetual colonial subjugation. However, despite the rhetoric, the Kennedy administration 'ended by pursuing similar policies of intervention', when confronted by civil war in the Congo.[4]

Writing in the 1970s, E. Jefferson Murphy and Harry Stein noted that 'Africa normally rates very low on the scale of American foreign policy priorities'.[5] They listed five areas of importance:

- Africa's ores, minerals, investment and trade opportunities.
- Africa's geostrategic importance.
- Africa's symbolic importance to African-Americans.
- Africa's increasingly important voting bloc in the United Nations.
- And 'America's position of world leadership' requires our moral and political response to Africa's conditions of underdevelopment.

In this set of stated interests the socio-humanitarian dimension of America's interests in Africa was expressed. These served to anticipate the American public's – especially the Afro-American community's – concern with anti-apartheid measures leveled against South Africa in the mid-1980s.

Robert Price depicts the attitudinal change from the Nixon/Ford to the Carter administrations, in which United States Secretary of State Henry Kissinger, representing the former, summarized United States policy as the 'prevention of the radicalization of Africa'.[6] The Carter Administration sought to depart from this characteristically Cold War posture, but the response to Soviet advances in Africa soon blurred the distinction between the Nixon and Carter approaches. Price argues that in fact the perceived security dangers for both Administrations were myths all along, and instead, the United States had significant economic interests in the region, but these were never really threatened.[7]

The early Reagan administration's posture on Africa's relevance to the United States was expressed by Under-Secretary of State for African Affairs, Chester Crocker. He itemized as follows:

- Africa as the source of minerals, some vital for defense industries.
- Oil, especially Nigeria's, which at times was our second largest supplier.
- The prospect of increasing the quantity of mutual trade.
- United States air and naval access, tracking stations, transmitter and relay facilities.[8]

A few years later, a United States State Department document itemized Africa's importance to the United States:

- With over 50 nations, Africa is a significant factor in multilateral politics.
- It contains major oil and mineral reserves.
- Trade and economic cooperation opportunities.
- Strategic location for communications and transport.

- Africa's conflicts involved external intervention.
- Africa's economic stagnation invites external manipulation.
- African-Americans' political interests in Africa.[9]

And the White House's annual statement *National Security Strategy of the United States* (1988) identified the need to maintain military access to United States facilities in support of our strategic interests; Africa's source of strategic minerals and market for United States exports; Africa's shores adjoining important sea lanes; and Africa's significance as a voting group in the United Nations and in other international organizations.[10]

That document also elaborated Soviet interests in the Horn; their support for the war in Angola and expansionist interests in southern Africa; Soviet arms supplies to Libya and the latter's aggression in sub-Saharan Africa; and identified South Africa's apartheid discrimination practices as the primary cause of instability in southern Africa. Economic concerns were viewed as being closely interrelated with political stability, and market-oriented economic reform was critical for growth and development. The United States would support regimes threatened by Soviet and Libyan incursions; the United States worked towards the peaceful resolution of conflicts, and promoted progress towards non-racial, representative democracy in South Africa. As for the military assistance programs, the United States would provide a reasonable level of support to friendly countries such as Kenya and Zaire and to those 'on the front lines of Soviet-supported aggression', such as Somalia. United States military training programs were seen as being valuable, and in this regard the United States offered Africans a coastal security program designed for the protection of fishing resources. Other measures included aid efforts, food and economic policy programs, and our attempt to work on the debt problems in Africa.

By 1990, the newer version of that annual assessment offered a much briefer reference to Africa but stressed our continued efforts to help liquidate 'all the Soviet/Cuban interventions in Africa left over from the 1970s'. This was a significant interpretation as it did not recognize any new or ongoing Soviet expansionist efforts and therefore signaled a turn in United States policy, which had been unmistakably reactive to Soviet Cold War initiatives for three decades.[11]

United States Interests in Africa

When summarizing broadly the various analytic or officially enunciated American interests on the African continent which have been expressed during the last four decades, three standard analytic categories of concerns may be identified:

Political Interests

Generally, the shifts over time in America's perceptions of the African continent reflect great initial apprehension of the importance of Africa's emerging state entities whose potential role in the Cold War was very uncertain. The continent's geostrategic location and physical resources were attributes that might be measured, but the human dimension was poorly assessed. In the midst of the Cold War, could the United States afford to let Soviet advances in Africa go unchallenged? The continent offered the Soviets fertile ground for adventurous probing in a region where mistakes could be made without lethal consequence. However, these excursions engendered America's reactive policies in Africa and generated the Soviet Union's power of initiative.

The following factors represented the major variables in the East–West competition for Africa's allegiance:

1. The addition of ideological allies to the Western system while conversely depriving the Soviet camp of potential adherents. Gradually the United States realized that Africa's pro-Western stance was an ingrained feature of a frustrated continent for which the Marxist operational paradigm offered little attraction.
2. The attempted demonstration by the United States that the Western conceptions of liberal democracy and market economic structures are universally valid, regardless of the greatly inferior level of development prevailing in those new Third World states.
3. Inculcating the new states with a Western lifestyle and popular culture by extending linguistic, cultural, religious, and above all, educational facilities which were to attract the younger African generations towards the West. The spin-off of the technical expertise, which our Peace Corpsmen provided in Africa, was to have them also serve as our cultural ambassadors.
4. Gaining the voting strength of the numerous African states in the United Nations on our side would come at a price, but one perceived to be a good investment. They were to help ward off the Soviet-centered Marxist empire in the vulnerable parts of the world.
5. America's black community would increasingly express concern about government policies towards Africa. This did in fact become evident in the well-organized leadership which blacks in and out of

government demonstrated with regard to the South African sanctions campaign, and the enthusiastic reception accorded to Nelson Mandela in 1990. African-Americans had engaged in some effective and highly visible domestic anti-apartheid activities aimed at weakening politically South Africa's white regime.

6.   Africa offered fertile ground for the expansion of America's idealistic aspirations, expressed most concisely in the concept of human rights. Rewards of aid, or punishments (by its deprivation) offered Africa as a demonstration forum in lieu of other world arenas being too sensitive or inopportune for such experiments. Another dimension of this, evidenced during the Carter administration, was the deliberate attempt to preclude viewing Africa in only globalist Cold War terms. The United States was to address Africa's unique developmental needs, which in turn would ensure the progressive improvement of their quality of life, and peripherally remain in the Western camp.

7.   And the United States saw the moral need to distance itself from South Africa's apartheid-based policies which posed anyone dealing with that government with a no-win moral situation.[12] Modest short-term economic gains could be had in South Africa, but at the cost of offending the rest of the continent. But although such economic advantage could be sacrificed, South Africa had been a valuable strategic ally in the attempt to neutralize Soviet and Cuban objectives on the continent.

Economic Interests

Where America's political interests in Africa were largely nebulous and *ad hoc* reactive, our commercial expectations were supremely naive. Official policy pointed to the expansionist prospects of economic gain – which would surely follow the investment of a bit of seed money. Advising the continent's inhabitants to adopt Western models of development was as easily offered, as it was irrelevant. Yet the vast sums poured into Somalia and Liberia demonstrated, if anything, an inverse relationship between aid and stability. Other economic factors defined our economic attitudes towards the African continent:

1.   Africa's minerals exports had been commercially well-established but this commodity received a boost when South Africa's 'strategic minerals' such as chrome, manganese, vanadium, and industrial diamonds were highlighted as having Cold War relevance. This concern formed the major argument of those in the United States who would ease the pressure on South Africa.

2. OPEC's strong stance on oil in the early 1970s vaulted Nigeria, and subsequently Angola, into prominence. As Nigeria's major oil customer, the United States soon experienced a large trade deficit with Nigeria, while in Angola, oil played a paradoxical role as Cuban troops competently protected private American-owned pumping facilities, although the United States actively backed Jonas Savimbi's UNITA movement, the government's major insurgent opponents.

3. Traditionally, one-third of United States trade with sub-Saharan Africa had been with South Africa; another third with Nigeria; and all other countries combined comprised the final third. Yet the combined total barely constituted less than 5 percent of our total foreign trade. Initially, we had expected that Africa's markets would expand rapidly, but better trade opportunities in Europe, the Middle East, and then in the Far East, soon overshadowed commercial opportunities in Africa. Though Africa's population constitutes 9 percent of the world's population, the entire continent generates only 1.2 percent of global gross product as of the mid-1990s. Recently issued World Bank data point to substantial drops from 1980 to 1990 in Africa's vital economic statistics concerning per capita output, consumption, investment, exports, food production, and poverty.[13] These data are still declining, and along with it, Africa's waning bargaining power in the world's highly competitive economic arena.

4. Africa's new states were initially perceived by America's private sector as offering potentially great investment opportunities. But due to the lack of physical, financial, and managerial infrastructures, and adequate market size, this did not materialize. Besides, the Europeans, as ex-colonial powers familiar with the region, had a decidedly greater advantage at the outset, while for Americans the investment opportunities elsewhere were simply far more attractive and much less volatile. Among the business community, it is well known that profits in Africa may be had, but their realization requires a greatly disproportionate effort in comparison to other global regions.

5. Africa remains an exporter of primary products – most of which cannot take a forceful stance against international market power. The terms of trade have hurt Africa's scant exports, and that continent remains vulnerable to the vicissitudes of the international energy markets. The American private sector prefers not to engage in the slow development process, hence the African continent is left to the public sector and to a variety of international organizations with aid programs – many of which demonstrate little return on their programs and hence they encounter inevitable donor fatigue.

Strategic Interests

Ironically, while Africa played only a very marginal role in the global Cold War strategic stakes, in reality this dimension generated the greatest amount of publicity and shaped the American public's image of the Cold War's primary adversaries battling it out by their respective surrogate forces. The surrogates did battle, but there was little prospect of a direct armed confrontation between the two principals in Africa or over an African issue. Nor were all of Africa's hot conflicts engendered by external forces. Oil is a much more visible force driving conflicts generated by Africans themselves. Nevertheless, a review of the strategic dimension of Africa's security environment elaborates the third category of America's interests on the continent:

1.   Africans have never posed a direct security threat to the continental United States, but since 1960 Africa has been a site of our strategic posturing for other proximate geographic areas, notably, the Mediterranean region and the Middle East, or against our major Cold War adversary, the Soviet alliance. During World War II, North Africa, the Horn region, and to a much lesser extent, West Africa, were the sites of numerous military engagements, and supporting activities such as transport, staging, or communications facilities.

2.   North-East Africa today retains modest strategic geographic advantages for security-related activities in the Mediterranean region, Israel, and the Middle East. The increasingly anarchic Horn region is precariously close to the Middle East's oil supplies. In addition, from East Africa, the Gulf region and Indian Ocean may be monitored, and support rendered to the Indian Ocean island, Diego Garcia – a useful British-controlled air base asset during the 1991 Gulf War – enhances the further strategic significance of eastern Africa.

3.   The South Atlantic zone may be monitored and serviced from southern Africa, which in effect means South Africa, as the other countries in the region are too weakly equipped to provide us with appropriate facilities. For the United States this has offered only marginal utility, but this advantage was more important for the British defense of the Falklands.

4.   Much was made of the vital strategic sea lanes surrounding Africa, especially around South Africa. This was to have concerned the shipment of oil and minerals, and in wartime, the movement by navies. But it had been conceded that this was more a concern to Europeans than to Americans, and that in time of war, should the Soviets have attempted interdiction of American ships in those distant waters, the United States

would no doubt have escalated an air war to a level where those ships could have played only a marginal, if not an irrelevant, role.

5.   Several countries in Africa offer a variety of security related advantages such as berthing rights, ship-repair facilities, airports, training, radar and communications facilities, and space-tracking stations. These are important, but as non-hostile activities, utilization of them can be negotiated in exchange for payment. These are also vulnerable facilities, as the governments with whom agreements are made, may not survive for long. Ethiopia previously, and subsequently Liberia, are examples of this problem.

6.   The most important, and activist dimension of our strategic interests in Africa, had concerned the periodic attempts to aid in the installation or support of a friendly government in power, and the neutralization of advances made by external interveners, namely the Cubans, Soviets, Libyans, and Iranians. This had led the United States in several complex involvements with various African governments dating back to the initial Congo operations in 1960. Several such involvements on behalf of our preferred head of government, have not attained lasting success, as evidenced by our vacillating diplomatic experiences in Sudan, Ethiopia, Somalia, Kenya, Liberia, and Zaire, to name a few.

7.   America was only a minor arms supplier to Africa in contrast to the Soviet Union; nor did the United States spend billions to establish control over African countries as had the Soviets in Angola and Ethiopia in conjunction with Cuban armed forces. At a lesser level, the United States did support the early days of President Doe in Liberia ($500 m) and Barre in Somalia ($700 m). By the late 1980s, the Soviets had recognized the futility of their energetic and costly strategic operations in Africa, having made no permanent strategic or diplomatic gains.[14] Overall, America's neutralization efforts of Soviet and Cuban activities have been modest in cost, and moderately successful.

Analytic Perspectives

Expressed as a more concise overview of America's perception of the African continent up to the end of the Cold War, we may identify certain characteristics regarding America's interests from the above-listed factors:

1.   As the African countries attained independence, America was initially poorly prepared to anticipate correctly the consequences of the expansion of the global system to include the new states of Africa.

2.   The United States perceived the new African states in terms of our Cold War strategic requirements and not in accordance with the

strengths and weaknesses of the independent, but fragile, entities which they were. Nor did the United States assess the prospect of building mutually advantageous and productive relationships with these entities. Africans would be easy prey to powerful external interests and, ultimately, Africa would offer little more than only a surrogate human battleground and fighting terrain – none of which addressed Africa's desperate need for basic development.

3.   We learned only reluctantly, as did the Soviets and external Third World revolutionaries, that Africa was not a receptive region for implementing externally derived ideologies and externally modeled political and economic institutions. Whether democracy, socialism, capitalism or variations of authoritarianism, such concepts played a very indeterminate role in Africa. In their orthodox forms, all foreign paradigms were equally inappropriate. In one noted analytic effort, having examined the relationship between ideology and positive socio-economic performance, Crawford Young offered his influential observation, 'My reading of the evidence does not lead to a single, unambiguous conclusion'.[15] America's current predisposition towards civilian rule and multiparty democracies, with effective courts and constitutional guarantees, may well be noble objectives, but Africa will first have to develop the social infrastructure and effective public institutions in order to support the necessary requirement of political legitimacy.

4.   The United States misassessed – as did Africans – the gravity of economic problems that would follow Africa's waves of independence. We emphasized excessively the need for Africa to adopt economic structures, policies, and practices, which approximated our own. The United States offered to be good economic partners, but Americans were oblivious to the vastly disproportionate bargaining power they possessed, which hardly enabled Africans to negotiate as equals. Yet, Africa had no realistic alternative to Europe and the United States as economic partners, because the Soviet Union offered little non-military aid, trade, and investments.

5.   Strategically, the United States soon regretted the expansion of the Cold War to Africa, however, we did not hesitate to pursue allies and advantages with which to neutralize Soviet and Cuban initiatives on the continent. Our own commitment was never large, nor did the African battleground rank as a vital geopolitical location. But it was Africans who paid the human cost of battles – and continue to do so – regardless of the nature of the external interests. America's own security interest would hardly be enhanced perceptibly if our preferred combatant in a war among Africans emerged victorious.

6.   The United States was a mostly reactive power to Soviet, Cuban, Libyan and other countries' initiatives in Africa. Much of United States' aid programs were determined not so much by humanitarian need, as by our global diplomatic and security requirements. This had the ironic consequence in that Soviets and other external interests were in a position to determine which African nations would gain America's benevolent attention in the face of responsive intervention.

7.   And despite the occasional attempts to extend America's own idealistic and humanitarian aspirations to Africa, these efforts were often thwarted by more immediate security requirements in other parts of the world. This had the effect of relegating Africa's human concerns to a distant secondary position to global, Cold War sociopolitical determinants or economic security requirements. This was amply demonstrated in 1990 by contrast when the United States, despite earlier massive financial aid infusions to the inept administration of Liberia's Samuel Doe, stood by as little more than a perplexed observer of that country's total disintegration, but at about the same time, preparations for a huge and expensive military operation were undertaken in the face of Iraq's aggression against tiny, but oil-rich, Kuwait.

This seemingly cynical appraisal of America's security-centered policies towards Africa is not meant to highlight a deliberate disdain for Africa's non-security related concerns. America's foreign policy with respect to most global regions had also been shaped by exigencies of the Cold War. Europe is our major ally and the location of a line that the Warsaw Pact's military machine was not to cross. Latin America is our backyard, and while some unruly neighbors were only reluctantly tolerated, all nations of that region were to respect our sensitivity concerning the unwelcome activities of their occasional Soviet guests. The Middle East's Islamic world has likewise been a potential revolutionary cauldron but its positive feature is that this region's inhabitants remain naturally averse to the Leninist brand of Marxism. And East Asia had experienced America's massive commitment to halt the spread of the Sino-Soviet empire – until the clear split between the Soviet Union and the People's Republic of China became evident, which indirectly, allowed the United States to depart ignobly from Vietnam.

Under the circumstance, viewing Africa today as having been but a peripheral extension of a global security network should not be surprising. Africans may resent such pragmatic treatment, but the United States easily countered by portraying the devastating revolutionary history of communism which surely would have shown great disrespect for

Africans' wishes altogether, were it not for America's benevolent neutralization of Soviet regional ambitions. Soviet involvement in Africa was not halted by a policy alteration of their revolutionary ambitions, but by domestic ideological and severe economic exigencies. As soon as the Soviet Union's General Secretary Mikhail Gorbachev had publicly announced his intention to begin a phased withdrawal from Afghanistan, the South Africans correctly perceived that this signaled the readiness for a withdrawal from other Soviet commitments as well-most probably from Angola in southern Africa. Otherwise, why would the Soviets have yielded their established commitment in Afghanistan – adjacent to the vital security zone of the Soviet Union – but retain the geographically much more distant commitment in Angola, which had never been explained? Whatever original formulations led the Soviets to work with the Cubans in order to secure a socialist ally in power in Angola, in view of the radical alteration of relations with, not only Afghanistan, but also Eastern Europe, it was evident that the Soviet Union's controversial and expensive presence in Africa was seen by Gorbachev as having been at best only a foolish venture.

In the current, early days of the post-containment era, we need to appreciate that recording contemporary history is much more difficult than to write the history of a previous generation. We lack comprehensive data; details of evolving operations; insight into the perceptions of policymakers; and the reminiscences of key participants in Africa's decolonization process. From the scant post-Cold War era evidence, however, certain developments may be considered as possible harbingers of the shape of things to come as Africa's global role undergoes redefinition.

In this new era, the United States demonstrates no evident radical reorientation in policy which hypothetically, might be expected were the defense and aid budgets to experience decimation. America's extant global interventionist agenda was illustrated in the case of the late-1989 invasion of Panama, commensurate with our history of relations with Latin America. Next, the rapid and massive allied military and diplomatic response to Iraq's invasion of Kuwait may have been enabled by the absence of a Soviet counter-response, the prospect of which would no doubt have had to be taken into account in a previous age. Then followed military engagements in Somalia, Haiti, and Bosnia, and humanitarian involvement or non-combat evacuation operations in Rwanda, Zaire, Liberia, Sierra Leone, Central African Republic, Tanzania, as well as several more supportive missions in Somalia.[16] These were all minor military operations, but they served to illustrate that armed conflicts, and therefore the need to maintain a wide range of military capabilities, had survived the termination of the Cold War. Although most such engagements tended to be at a very low level of intensity, characterized as 'Military Operations Other Than War', the frequent calls for America's military projection capability will ensure that the United States will encounter a steady stream of challenges to regional stability. Africa, the Middle East, and South Asia are currently the world's most unstable regions, with Africa's armed conflicts being rated the least dangerous to global peace. A comprehensive agenda of American military involvement on the African continent will not be formulated on behalf of America's core security concerns. With the possible exception of North and North–East Africa, America's active military-supported role in Africa will be conducted mostly for humanitarian considerations – and not without limits.

These introductory comments have focused heavily on security issues, as this has been the most evident Cold War context for understanding America's perplexity with the emergence of many new states in a volatile region scarcely known to Americans. The major Cold War diplomatic morass concerned the very sensitive relations with America's primary source of security challenge: the Soviet Union. It has therefore been argued that analytically, Africa has offered little, if any, advantage for advancing our core global security interests in the past. However, the United States officially operated from a position of great uncertainty regarding the African continent. Currently in line with the termination of the Cold War, and with a questionable record of relations with the Third World, we may reasonably expect the United States to disengage from military interventionism in today's marginalized Africa and to limit the frequency of humanitarian involvement as well. Without a credible new opponent seeking to exert wide controls over Africa, that continent may be abandoned by the US and left to its own devices in the painful pursuit of domestic political consolidation and legitimacy.

Notes

1. William J. Foltz 'Africa in Great-Power Strategy', in William J. Foltz and Henry S. Bienen (eds) (1985) *Arms and the African: Military Influences on Africa's International Relations* (New Haven: Yale University Press), 2.
2. David A. Dickson (1985) *United States Foreign Policy Towards Sub-Saharan Africa* (Lanham, MD: University Press of America), 5.
3. Melvin Gurtov (1974) *The United States Against the World* (New York: Praeger), 43.
4. Ibid., 45.

5. E. Jefferson Murphy and Howard Stein (1973) *Teaching Africa Today* (New York: Citation), 200–2.
6. Robert M. Price (1978) 'US Foreign Policy in Sub-Saharan Africa: National Interest and Global Strategy', *Policy Papers in International Affairs* (Berkeley, University of California: Institute of International Studies), 59.
7. Ibid., 59–60.
8. Chester A. Crocker (5 October 1981) 'US Interests in Africa, Current Policy No. 330' (Washington, DC: United States Department of State), 1.
9. United States Department of State, (December 1985) 'Sub-Saharan Africa and the United States', *Discussion Paper*, 20.
10. The White House (January 1988) *National Security Strategy of the United States* (Washington, DC), 32–3.
11. The White House (March 1990) *National Security Strategy of the United States* (Washington, DC), 13.
12. The moral-economic interplay was covered in an article by Karl P. Magyar (1985) 'The American Disinvestment in South Africa Debate: Short-term Morality vs. Long-term Economic Development', *International Affairs Bulletin*, 9(1).
13. Data compiled by Donald L. Sparks (1997) 'Economic Trends in *Africa South of the Sahara*, 1997', in Africa South of the Sahara (London: Europa Publications), 11.
14. Donald Rothchild and Caroline Hartzell (1992) 'The Case of Angola: Four Power Intervention and Engagement', in Ariel E. Levite, Bruce W. Jentelson, and Larry Berman (eds) (1992) *Foreign Military Intervention: The Dynamics of Protracted Conflict* (New York: Columbia University Press), 165–77.
15. Crawford Young (1982) *Ideology and Development in Africa* (New Haven: Yale University Press), 324.
16. Institute for National Strategic Studies (1997) *Strategic Assessment 1997* (Washington, DC: National Defense University), 165.

1

The United States and North Africa: Sustained Strategic Interests

Mohamad Z. Yakan

The end of the Cold War has had a profound impact on United States foreign policy towards many countries and regions of the world. One of its major casualties was the global containment policy that was pursued by the West in the post-World War II era. The end of the Cold War put a halt to this policy, leading to major shifts in American strategic concerns. As a result, many political priorities of the Cold War era lost much of their significance, a factor which was sufficient to signal the start of a new stage in American foreign policy with new and different challenges. The stage is still evolving and it will take a while before its guidelines and frameworks consolidate and become clear. To borrow James Schlesinger's words:

> To this point the record is, at best, mixed. A plethora of foreign policy objectives has been put forward, as if all could be successfully and simultaneously pursued. We are urged to pursue democracy and all its procedures, human rights, civil liberties, equality before the law, protection of minorities, self-determination, an orderly world, international law, economic growth, free markets, privatization, free trade, limits on environmental degradation, curtailment of the arms trade, prevention of the spread of advanced weapons, etc. . . . [Some] of these objectives are flatly in conflict and that all require the careful examination of trade-offs. Moreover, striking little attention has been paid to the relation between means and ends. Individual tools are assumed to achieve multiple objectives – with little heed paid to their inherent limitations.[1]

In Schlesinger's view: 'With so many conflicting objectives and with an inability to focus those means appropriate for achieving a limited set of objectives, now foreign policy is likely to be shaped by a capricious flow of events – rather than defined guideposts and a careful plan.'[2]

The absence of clear and well-defined guidelines for the post-Cold War era explains the disarray that is being witnessed in the discipline of American foreign policy. Without clear and well-defined guideposts, studies on foreign policy devolve into mere speculation. The confusion is quite apparent in studies on American–African relations in general, and American–North African relations in particular. Some studies describe US foreign policy towards Africa as one of retreat.[3] Others portray it as one of cynical disengagement.[4] A few questioned whether Africa is falling off the map, and others call for remembering Africa.[5] All these studies, however, took for granted the fact that Africa is a continent of 54 independent states, each of which presents varied interests to the United States. Some African states have been and continue to be central to US concerns. Others have been and continue to be low on the scale of American interests.

Broadly speaking, these observations suggest that apart from general humanitarian and democratization policies, the US does not support an even or uniform policy towards African states in general, especially towards North African. That this is the case is quite evident in present American–North African relations. American–Egyptian relations, for example, are very friendly. Egypt is a recipient of over two billion aid dollars annually from the US, whereas Libya is considered a pariah state because, among other reasons, it declined to hand over suspects in the bombing of the Pan Am flight 103 over Scotland to be tried in the United Kingdom or the United States. At this juncture, American–Tunisian relations are congenial, but not as solid as those between the US and Morocco. Relations with Algeria are unclear and colored with great reservations. But at the same time they are supportive of the present Algerian regime which is committed to combating an Islamic fundamentalist take over of the country. Militant Islamic movements are a cause of alarm to the US, and hence are looked upon with great concern by American foreign policy decision makers.

These points suggest that the US does not have one policy that is uniformly applicable to North African states. A review of its foreign policy towards North Africa and its constituent states will not only confirm this notion, but also will show that the United States' policies towards North Africa have lacked consistency. Additionally, it will show that since the start of the Second World War, American–North African relations have become central to the national security objectives of the United States. Unlike most sub-Saharan African states, several North

African countries have become essential and strategically significant to the United States.

The end of the Cold War did not change the importance of North African states in American foreign policy. Rather, it confirmed the view that as long as the US is committed to the defense of Europe, the protection of the sources of oil in the Middle East and its free flow to the West, and the freedom of trade and access to markets, the US will continue to view North Africa as a region of crucial importance.

Prior to addressing these contentions, it is necessary that a brief historical overview of American–North African relations be offered. Second, it is also important to consider the geopolitical features of North African states, or what amounts to the special claims of North African states on American high (security) politics. A third concern of this chapter relates to American–North African relations in the twenty-first century – that is, the prospects for these relations in light of the developments generated by the end of the Cold War.

American–North African Relations: A Historical Overview

The study of American foreign policy can be broken down into four major phases. The first phase spanned from the Declaration of Independence on 4 July 1776, to the acquisition of Alaska in 1890, the year that marked the consolidation of the US in what is its present territory, to the start of World War I. This phase was an essentially formative one. The second stage was that of the First and Second World Wars period, US involvement in these wars, and the establishment of the League of Nations and subsequently, the United Nations. The third phase was that of the post-Second World War and the Cold War period. This phase ended with the collapse of the Soviet Union, and signalling the start of a fourth phase, namely that of the post-Cold War era.

It was during the formative first phase that US foreign policy started to take shape. Its principles focused on consolidating the country's independence, cultivating friendships with other independent states around the world, fostering the military capabilities of its armed forces, expanding its territorial space, developing trade relations with other countries, and protecting the commercial activities of its citizens abroad. The realization of these objectives was obstructed by European rivalries, persistent struggle between European powers for international hegemony, recurrent interferences with freedom of navigation, and acts of piracy on the high seas.

Broadly speaking, American foreign policy in this formative phase was highly pragmatic. It rested on an admixture of both political realism and idealism. It was realistic in the sense that it was based on an awareness of the threats posed by European colonial powers to American interests and the need to address them through self-reliance, as well as through measures conducive to the protection of American interests, including the territorial and commercial interests of the United States. At the same time, it was idealistic in the sense that it rested on the belief in the rights of people to determine their own destinies.

This admixture was evident in the US preparedness to address whatever threatened or appeared to threaten its territorial integrity or independence, as well as in its advocacy of policies that called for respecting freedom of shipping on the high seas, promoting freedom of trade and freedom of access to markets, and the right of foreign-ruled countries to self-determination and independence. These policies, however, were general policies applicable to all countries and regions of the world, including Africa. Because US interest and influence beyond the Southern Hemisphere and the Pacific region were not so developed, these policies expressed general moral, economic, and political attitudes and hence, with a few exceptions, did not bear 'any extensive practical implications.'[6]

American–North African relations were initiated during this formative phase, at which time the leaders of the American colonies sought recognition and financial and military assistance as independent states, especially from England's traditional enemies.[7] American foreign policy at this stage focused on cultivating the friendship of those North African states which were independent or semi-independent, namely Algeria, Egypt, Libya, Morocco, and Tunisia. Excepting Morocco, all North African states were nominally under Ottoman sovereignty. The recognition of these states was deemed important for consolidating the international status of the United States, developing markets where American products could be sold and raw materials could be obtained, and protecting American commercial interests.

Of all the states contacted, Morocco was the first to respond to the United States' quest for recognition and markets. In 1777, for example, Morocco recognized the government of the US, and in 1787, 'the two countries signed a treaty of peace and friendship and established commercial relations. Maintained in its essential provisions to the present day, the Treaty of Marrakech (Marrakesh), is the longest unbroken agreement of its kind in American diplomatic history.'[8] In addition, the US resolved outstanding issues resulting from the seizure of

American vessels in the Mediterranean by Barbary ships. Furthermore, under the terms of the agreement, the US agreed to pay Morocco an annual tribute of US $10,000.[9]

Four years later, an American consulate was opened in Tangier (Tanja), a famous Moroccan port. The compound which was acquired for the residency of the American consul in Tangier is considered the oldest piece of American public property abroad.[10] In 1836, the Treaty of Marrakech was supplemented by another treaty, that of Meknès (Meknās) in which Morocco formalized the most-favored-nation status granted to the US in 1824. Under the provisions of that treaty, Morocco also 'accorded extraterritorial privileges to American citizens, granting them capitulatory rights to be tried only by courts established at United States consulates'.[11] The Treaties of Marrakech and Meknès continued to govern Moroccan–American relations until 1911, at which time Morocco was partitioned by France and Spain and fell to de facto French and Spanish rule. As a result, direct relations between US and Morocco ceased. They did not resume until the latter's independence and reunification in 1956.[12]

The next North African territory to develop relations with the US was Algeria.[13] According to Nelson D. Harold, prior to US independence, American merchant ships frequented the Mediterranean, but under British protection. 'No longer covered by British tribute payments after the American revolution,' in his view, these ships were frequently seized with their cargoes and seamen. In his view also, the US had no choice but to conclude 'a treaty with the [Dey] of Algiers in 1797, guaranteeing payment of a tribute amounting to US $10 million over a 12-year period in return for a promise that Algerian corsairs would not molest American shipping'.[14]

Seizing the opportunity of a war between Algiers on the one side, and Spain, the Netherlands, Prussia, Denmark, Russia, and Naples on the other, the US Congress authorized in March 1815 a naval action against all Barbary states. According to Harold, 'Commodore Stephen Decatur [1779–1820] was dispatched with a squadron of 10 warships to ensure the safety of American shipping in the Mediterranean and to force an end to the payment of tribute.'[15] Commodore Decatur successfully captured several Algerian vessels with their sailors, bombarded Algiers, and more importantly, secured a new agreement with the Dey of Algiers. Under the terms of this agreement, the Dey 'agreed to discontinue demands for tribute, pay reparations for damage to American property, release American prisoners without ransom, and prohibit further interference with US trade by Algerian corsairs'.[16] In addition, the Dey agreed to place the US 'on the footing of most favored nation'.[17] Thereafter, the Algerian–American treaty of 1815 governed the United States' relations with the Dey of Algiers until 1830 when the latter country was conquered by France, and subsequently declared a department of France. As a result, direct relations between Algeria and the US ceased until Algeria's independence on 5 July 1962.

The third North African country with which the US developed contacts was Tunisia, then ruled by beys under nominal Ottoman sovereignty. In 1799, the US government signed 'a pact of friendship and trade with the Bey of Tunis, who guaranteed the inviolability of "American persons and their goods," in return for payment by the US of an annual tribute'.[18] According to Talcot W. Seelye, the pact was ill-received in the US 'where the slogan appeared, "Millions for defense, but not one cent for tribute". Yet, not having much of a navy in the late 18th century, the US was in a bind.'[19] The tribute was considered the only means at those times to protect the newly built Yankee clippers against assaults of the Barbary states' navy which commanded control over many portions of the Mediterranean Sea, especially its eastern portion. In 1805, however, the Tunisian Bey was pursuaded to forgo the tribute requirement. Instead of the tribute, the US donated the *Franklin*, a ship armed with several canons, to the Bey.[20] According to Seelye, thereafter, an affinity began to develop between the two countries, 'growing out of mutual opposition to the European powers and Tunisian admiration for American technical and maritime skills'.[21]

During the early part of the nineteenth century, the US expanded its relations with Tunisia and opened an American Consulate in Tunis. William Eaton and David Porter Heap were among the most noted American consuls who served in Tunis. The former's fame is derived from his organization of 'a force to depose the neighboring Bey of Tripoli – an operation, commemorated in the Marine anthem "to the shores of Tripoli"', in what is presently modern Libya.[22] The latter's celebrity is attributed to his donation of 'a large block of marble from the ruins of ancient Carthage ... for the Washington monument'.[23]

Tunisian–American relations continued to be governed by the pact of 1799 until the conquest of Tunisia by France in 1881, and the placement of the country under French protection. Thereafter, American–Tunisian relations were handled through France. No direct relations between the two countries were resumed until Tunisia's independence on 20 March 1956.

The fourth country to develop relations with the US was Tripoli (Tripolitania), the western province of what is presently modern Libya.

These relations were from the outset hostile in nature. They started when the Pasha of Tripoli demanded the payment of a higher tribute than was stipulated in the American–Tripolitan treaties of 1786–99.[24] His demand, however, was rebuffed, and moreover the US resolved to send expeditions against Tripoli, which resumed harassment of American commercial vessels in 1801. Commodore Edward Preble's (1761–1807) command assembled at Gibraltar in 1803. As a result of naval engagements in the Mediterranean, the US frigate *Philadelphia* fell into Tripolitan hands after losing 300 of its crew.[25] In 1804, another naval expedition led by officer Stephen Decatur was dispatched into the harbor of Tripoli (Tripolitania). The purpose of the mission was to burn the US frigate *Philadelphia*. Decatur 'succeeded in this objective and made his escape under fire with only one man wounded'.[26] This mission was followed up by a land attack in 1805. The latter expedition was led by the US Consul-General William Eaton (1764–1811) and a claimant to rule of Tripoli. According to Jane S. Nickerson, the force was assembled in Egypt 'and after a fine march across the desert attacked and captured Derna, in concert with the fleet'.[27] The war, however, ended before Eaton's arrival in Tripoli. A truce between the Pasha and the US was established. Subsequently, a treaty of peace was signed with the Pasha of Tripoli, 'which provided that no further tribute should be demanded and that American vessels should be forever free of his exactions'.[28]

In 1816, Decatur commanded with success another naval operation against Tripoli. This incident, however, did not halt Tripolitan hostile operations against American commercial vessels, and hence, hostilities between the two countries continued until the fall of Tripoli and Cyrenaica (the eastern province of modern Libya) to Italian rule, starting in 1911. No formal American–Libyan relations were conducted again until the end of the Second World War.[29]

Regarding American–Egyptian relations, historical evidence suggests that they were not limited only to commercial ventures, missionary, and educational activities. They also covered the area of shipbuilding. According to Harry N. Howard, 'American shipbuilders had played a major role in reconstructing the Ottoman fleet destroyed by European powers at the battle of Navarino in 1827'.[30] Alexandria was one of the most important Ottoman naval bases, and hence was involved in the reconstruction effort. More importantly, history suggests the existence of cooperation between the two countries in the military field.

Americans were also involved in Mehmet Ali's rise to power after the expulsion of the French from Egypt by a joint Anglo-Turkish force in 1801. According to Clarence Clendenen, Robert Collins, and Peter

Duignan, 'In 1820 and 1821 George Bethune English, native of Massachusetts, graduate of Harvard, [and] former officer of the United States Marine Corps, [served as] chief of artillery in an Egyptian expedition' that Mehmet Ali 'dispatched to destroy the remnants of the Mamelukes'.[31] Mehmet Ali's successors, the Khedives, continued to cooperate with the United States.[32] In the early 1870s, Khedive Ismail for example, engaged the services of more than 30 top American military experts for the purpose of organizing and training his forces along modern lines. Their employment was facilitated by the American government, specifically by William T. Sherman, the Commanding General of the US Army, who gave these experts his personal recommendation and allowed them 'to take indefinite leave of absence for the sake of the practical experience they would gain in Egypt'.[33] The mission of the American team ended sometime after Great Britain's occupation of Egypt in 1882.[34] With the placement of Egypt under British protection in 1914, direct American–Egyptian relations virtually ceased until Egypt's independence in 1922.[35]

American Protestant missionaries started their activities in Lebanon in 1823 and eventually found their way to Egypt where they established a network of schools, colleges, and hospitals. One of their landmarks in Egypt is the American University of Cairo (AUC) which was established in 1919.[36]

In light of this review, United States–North African relations during the formative phase of American foreign policy were limited. They entailed limited commercial activities in generally inhospitable markets. Military force in defense of commercial interests was utilized on numerous occasions.

American Foreign Policy Towards North Africa From World War I to the Outbreak of the Cold War

The First World War signaled the start of the second phase of American foreign policy. This phase had a strong positive impact on American–North African relations. United States' prestige was enhanced by President Woodrow Wilson's principle of self-determination. The principle was well received not only by the peoples of North Africa but also by others under foreign rule.

The US Senate rejection of the Treaty of Versailles in which the League of Nations order was incorporated, however, signaled a period of American lack of interest in European international affairs, including those towards North Africa. According to John A. Garraty and Peter Gay,

'Disillusioned with the fruits of war, frightened by the spread of Bolshevism in Europe, the United States in 1919 turned in on itself.'[37] In their view, 'In the frenzy of the Red Scare [the US] sought to exorcise all European ideas ... '[38] This 'indifference' lingered until the outbreak of the Second World War.

During the Second World War, the US was a party to the Middle East Supply Centre, which was headquartered in Cairo. Moreover, US troops took part in the North Africa campaign and acquired military bases from the French government in Morocco, utilized French military bases in Algeria and Tunisia, used British military bases in Egypt, and together with Great Britain established military bases in Libya. In spite of these activities, however, the United States' political role in North Africa continued to be decidedly secondary. On the whole, the US rarely tried 'to meddle' in what it considered to be European concerns.[39]

Soon after the close of the Second World War, the US started to modify its former aloofness. The first signal of this change took place on 6 April 1946, when President Harry Truman warned of the weaknesses and instabilities in the whole Middle East area, including North Africa, and more importantly, highlighted the region's vital importance to the West. Still, President Truman showed unwillingness to address the issues of the area unilaterally, and more specifically, outside the framework of the newly established United Nations.[40] With the eruption of the Cold War, however, the US altered its detached position with regard to political, military, and economic responsibilities in North Africa. Consequently, the containment of Soviet expansion and influence became the criteria for determining the foreign policy to be applied to North African states.

American Foreign Policy Towards North Africa During the Cold War Era

The US now faced a dual problem: the exigencies of holding on to its European allies on the one hand, and the aspiration of North African countries for independence from America's European Allies on the other. A third major concern of American foreign policy was, in Lawrence G. Potter's words, 'to exclude the Soviet Union and secure American lines of communication in the Mediterranean'.[41] Understandably, a categorical position with regard to either problem was regarded critical to the United States' new global interests. The containment of the Soviet Union depended not only on maintaining a good relationship with Western European allies, but also on developing a productive relationship with the peoples and governments of newly independent countries. The latter requirement called on the US to reaffirm a traditional American principle, namely that of the rights of peoples to independence from foreign rule. The US opted to resolve these issues as they arose, but without clear decisiveness. Moreover, it tried to moderate the adverse effects of this indecisiveness on newly independent countries or countries struggling for independence through miscellaneous economic and financial aid programs. At the United Nations, for example, the US paradoxically, according to Polk, tried to water down proposals of the anticolonial bloc . . . and then vote with it against the European colonial powers, thus maintaining [its] traditional principles, yet not antagonizing European countries as much as [it] might have.[42]

This indecisiveness colored United States–North African relations throughout the post-World War II period, and in certain cases, it proved to be detrimental to American national interests. First, it left great latitude to European colonial powers in shaping the post-World War II era. Second, it undermined the US popularity in developing countries, and helped transform the image of the US from that of an anticolonial and principled power to that of a power motivated by expediency and one that was complacent to the aspirations of countries that were struggling to achieve their independence from colonial rule. Third, it indirectly helped promote the Soviet Union as a 'reliable friend' in many developing countries. The Soviet Union seized the opportunity of American indecisiveness with regard to issues of colonialism to promote its relations with newly independent states and liberation movements, including those of North Africa.[43]

Concerned about Soviet strategic interests in, and threats to, the Middle East and North Africa, the US initiated a new and unprecedented course of political involvement in the region's affairs. Alan R. Taylor notes however, that the new course was 'without a clearly defined policy toward the area'.[44] The containment of the Soviet threat was made the highest priority in the US policy towards the Middle East and North Africa.[45] To this end, in 1951, the US proposed the establishment of a regional alliance, what became the Central Treaty Organization (CENTO), to serve as a link between NATO and SEATO with Egypt as the 'linchpin of the system'.[46] However, the policy had almost exclusive focus on the Cold War with little regard to the interests of the countries that were targeted to support it. Moreover, it depended on the UK to rally support for it in the region. Furthermore, it underestimated the seriousness of the Arab–Israeli conflict, as well as traditional factionalisms in Arab politics.

Egypt's position at that juncture was not against being associated with the United States' containment policy, but its leaders preferred that this be done through an Arab regional organization under Egypt's own leadership. Encouraged by the United Kingdom, Iraq, Egypt's rival for Arab leadership, took the initiative of forming the Baghdad pact, precipitating a problem in the Middle East, and eventually transforming the area into a Cold War political battleground. Faced with this *fait accompli*, in 1955, the US decided to withdraw its offer to help build the Egyptian Aswan Dam project, leading the Egyptian government to nationalize the Suez Canal Company. This led Egyptian leadership to accept Soviet offers not only to finance the dam, but also to provide Egypt with military weapons.

Egypt's nationalization of the Suez Canal Company and its acceptance of Soviet offers of help served to transform the country into a pariah state in the eyes of the American government. But at the same time, the incident enhanced Egypt's leadership role among Arab masses throughout the Arab world. As a consequence, American–Egyptian relations deteriorated severely, resulting in a severance of diplomatic relations in the late 1960s. These relations continued to be colored with tension and hostility until the death of Egyptian President Gamal Abdul Nasser on 28 September 1970.

It was only upon Anwar Sadat's succession to Egypt's presidency that the grounds for improved American–Egyptian relations were set. Disillusioned with Soviet support and equipment, Sadat had decided to expel most Soviet military advisers in July 1972, and subsequently asked the People's Assembly to abrogate the Egyptian–Soviet Treaty of Friendship and Cooperation in March 1976. In June 1974, American President Richard Nixon visited Egypt, and in October and November 1975, Egyptian President Sadat reciprocated with a ten-day visit to the United States. Since then, American–Egyptian relations have witnessed steady expansion and improvement. The shift in Egyptian foreign relations prompted the US government to provide American arms to Egypt, as well as to add Egypt to the list of major recipients of American economic assistance.[47]

With Egypt's concluding peace with Israel in 1979, American–Egyptian relations deepened to an unprecedented level, making Egypt the most important US ally throughout the whole Arab world. More importantly, Egypt became the second largest recipient (after Israel) of American economic and military assistance. Between 1975 and 1996, according to Denis J. Sullivan, 'Egypt has received some $21 billion in economic aid from the United States plus over $25 billion in military aid.'[48] Currently, the 'US aid program in Egypt is the largest such program in the world'.[49]

The improved American–Egyptian relations were also evidenced in increased American military presence in Egypt, joint American–Egyptian military exercises, and Egypt's agreement to allow US Central Command (formerly US Rapid Deployment Force) to use Egyptian military facilities.[50] By 1989, for example, the number of American military personnel in Egypt rose to 1,200, and beginning the same year, both countries 'had held joint military exercises every other year under the name of Operation BRIGHT STAR ... In alternate years, [they] also held combined air and sea exercises.'[51] In 1989 also, Egypt agreed to avail the US Central Command access to the Ras Banas base on the Red Sea 'in the event that a friendly Arab nation needed help in repelling an armed attack'.[52] In addition, Egypt agreed to provide the US Navy with facilities on its territory, a strategic gain that the US had not had since the end of the Second World War.[53]

More recently, the improvement in American–Egyptian relations was epitomized during the Gulf War, as well as in the continued collaboration of both countries in supporting the Middle East peace process. Egypt was the first North African Arab country to support the American-led UN coalition against Iraq's invasion of Kuwait in August 1991, not only in public statements and rallying Arab support to the American-led coalition, but also in dispatching a 35,000 member force – the third largest after those of the US and Saudi Arabia – to the Gulf. Moreover, Egypt is one of the few Arab countries that has continued to facilitate diplomatic contacts between the parties of the Arab–Israeli conflict. American–Egyptian relations 'became particularly close during 1991–2, when the United States forgave $7 billion of Egypt's debt'.[54]

American–Egyptian cooperation, however, transcends the political, economic, and military domains. It also covers the educational, social, cultural, and scientific spheres. In the economic sphere, American direct foreign investment comprises Egypt's main source of external investment capital.[55]

Another example of indecisiveness was evident in American–Libyan relations. For instance, when the Great Powers met in 1945 to determine the fate of Libya, a former Italian colony, American Secretary of State James Byrnes proposed placing Libya under a UN trusteeship. Other Allied powers had different views. The Soviet Union claimed the right to become the trustee itself; France wanted to return Libya to Italian rule; Britain refused. In the ensuing months, as the Cold War began to be codified, the British Labor government, curiously, adopted the French position, and agreed for a renewal of Italian rule.[56]

The US, the leading power in the world, however, did not insist on its position, or what seemed to be the right thing to do. The impasse was finally overcome when the foreign ministers of the Allied powers decided that a Four Power commission be established to ascertain the wishes of the Libyan people. According to Polk:

> Libyans expressed themselves quite clearly but the Commissioners had their reports prepared before they visited the disputed area. The report of the Commissioners was a jumble of contradictions. Each Commissioner left his ears in his own capital. The dispute was turned over to the United Nations General Assembly. Finally, as all compromise measures failed, the Political Committee of the General Assembly decided that Libya should become independent no later than January 1, 1952.[57]

On 24 December 1951, the United Kingdom of Libya under the rule of Sayyid Mohammed Idris al Senussi was proclaimed.

Following Libya's independence, both the US and the UK negotiated agreements with Libya to continue their leases of their respective military bases, which they had established during the Second World War period. The lease of the United States' Wheelus Field outside Tripoli was entered at least until 1971, whereas that of the UK's near Tobruk went until 1973.

Though it dominated an important strategic position relative to the eastern Mediterranean Sea, Libya appeared marginal in American foreign policy. This position was secured further when Libya's king cultivated closer ties with Egypt, then at odds with Western powers, and adopted a neutral position in the East–West conflict. This marginality continued until the mid-1950s, at which time oil was discovered in Libya by a number of Western companies, including ESSO. This development attracted American business and investment into the country, and began to reverse the country's marginality. However, a *coup d'état* staged by Mu'ammar al-Qadhafi in September 1969 halted this process. The new regime proclaimed a Nasserist, non-aligned, and anti-Western agenda. In addition, it forged close military and trading ties with the Soviet Union, and at times, appeared to be furthering the Soviet agenda in the Middle East, and East and North Africa. Furthermore, it appeared to be backing and training revolutionary, terrorist, and opposition groups against friendly regimes not only in North Africa, but also in several other parts of the world. After Qadhafi's September revolution in 1969, American–Libyan relations became hostile, and remain so until this day.

These hostile American–Libyan relations were exemplified in the listing of the Libyan government as one of the world's sponsors of international terrorism, US prohibition of travel to Libya, US punitive campaigns against Libyan targets during the Reagan Administration, US imposition of a trade embargo on Libya, American training of Libyan dissidents in adjacent states, and in US support of continued UN sanctions against Libya.[58]

In April 1986, the US blamed Libya for the terrorist bomb that destroyed a discotheque in West Berlin, killing two persons, among them a US soldier, and wounding 204 others. In response, US President Reagan ordered a punitive air attack against the Libyan cities of Tripoli and Benghazi.[59]

The decisive blow came in December 1988, when 'a US jetliner, Pan Am 103, exploded in air while over Lockerbie, Scotland, killing all 258 people aboard and 58 on the ground'.[60] After extensive investigations by the US and other Western allies, fingers pointed at two Libyans, presumably working for the Libyan secret service. When Libya, in April 1992, refused to honor an American-sponsored resolution that called on it to hand over the suspects, the UN Security Council 'imposed economic sanctions and an international ban on air traffic and arms sales to Libya'.[61] The sanctions are still in force.[62]

American indecisiveness was also evident during the Algerian, Moroccan, and Tunisian struggle for the independence of their countries. In these crises, traditional American anticolonialism 'lost much of its earlier sharp cutting edge as the American position in the world'.[63] Instead, the US opted for a policy that appeared, at least in the eyes of Algerians, Moroccans, and Tunisians, as supportive of colonial rule, and at times as no different from those of colonial powers. The outlines of this new policy were laid out in a basic statement by Henry Byroade, the deputy assistant secretary of state in November 1953. Highlighting the hazards of hasty decisions with regard to ending of colonial ties, Byroade 'insisted that independence must have solid foundations if it is to last and that it is neither a cure-all nor necessarily a path to the ultimate objective of individual well-being'.[64] In this regard, he pointed out that 'the gravest and most constant complications arise from [US] relations with the European colonial powers. Here he pleaded for frankness, stating: "Let us be frank in recognizing our stake in the strength and stability of certain European nations which exercise influence in the dependent areas ... [whose governments] will probably represent, for many years to come, the major source of free-world defensive power outside our own. We cannot blindly disregard their side of the colonial question without injury to our own security".'⁶⁵

Byroade was frank in pointing out which side the US had chosen and why. Mason Sears, US representative to the United Nations Trusteeship Council in the 1950s, however, was more candid than Byroade in this respect. In his view, 'Because of the Cold War, [Secretary of State Dulles] decided to subordinate the promotion of African freedom to what he considered the overriding necessity to support ... NATO allies and their colonial policies.'⁶⁶

Clearly, this policy was not helpful in winning either the support of the liberation movements that were roaming in several North African states, or that of newly independent states whose governments were committed to helping these movements. The US opted for policies that would not weaken nor alienate its European colonial allies. In so doing, it indirectly encouraged liberation movements to seek help from US adversaries, including the Soviet Union, other communist states, and certain non-aligned nations.

As a consequence of this policy, the US resolved not to help Algerians, Moroccans, and Tunisians in their struggle to achieve the independence of their countries and to take stands that were supportive of colonial powers. Hence, on their independence, Algerians, Moroccans, and Tunisians felt that they had succeeded in their struggle despite American support of their former colonizers. They felt that they owed much to the Soviet Union and non-aligned nations, but nothing to the United States. By the time succeeding American administrations realized the folly of the North African policies of the Truman Administration, it was too late to halt the spread of Soviet influence in the countries of this region.

Insofar as Tunisia is concerned, American indecisiveness was evidenced during the country's struggle for independence in general, and by the French–Tunisian crisis of the 1950's. When France agreed to grant independence to Tunisia in March 1956, it insisted on retaining the naval base at Bizerte. As Tunisian demands that the base be relinquished were ignored, the issue dragged on and galvanized Afro-Asian bloc as well as Soviet support to Tunisia's position within and outside the UN arena.⁶⁷ The crisis provoked a series of condemnations from many countries, including some NATO members such as Denmark, Norway, and Turkey.⁶⁸

Tunisia's assertive measures, coupled with France's growing international isolation especially in the countries of the Third World, prompted France finally to agree to the relinquishment of the base to Tunisia.⁶⁹

Only then did the US change course. The US resolved to help Tunisia overcome its serious economic difficulties through grants from the US Agency for International Development (US AID) and the International Finance Corporation (IFC) of the World Bank. Reportedly, the US also resolved 'to sell a limited quantity of small arms to the infant Tunisian army, rather than risk Bourguiba turning to the Russians – as Nasser had done with the Aswan dam'.[70]

Broadly speaking, however, the grounds for improved American–Tunisian relations were set in the latter part of the 1950s, following the visit of Vice-President Richard Nixon to Tunis in March 1957. The visit came three months after the establishment of the Eisenhower Doctrine, which offered American military and economic aid to any Middle Eastern country endeavoring to withstand direct or indirect communist threats. The visit was fruitful and culminated in the conclusion of an agreement that provided for extending American economic, technical, and military assistance to Tunisia.[71] More importantly, Tunisia agreed to allow the Sixth Fleet, including American nuclear warships, to visit its ports.[72] Henceforth, the US government also supported Tunisia's endeavors to upgrade its armed forces, especially in the face of mounting Libyan threats to the country's integrity.[73] In order to help Tunisia finance its arms purchases, the US increased its grant assistance through the Military Assistance Program (MAP), the Economic Support Fund (ESF), and the International Military Education and Training Program (IMETP).[74] In fiscal years 1985 and 1986, the US Congress approved a $20 million economic aid package to Tunisia in recognition of its 'worsening economic situation and its importance as a moderating factor in the Middle East peace process'.[75] Between 1957 and 1985, total American aid to Tunisia amounted to about $1 billion. The improved relations between the US and Tunisia were also evidenced in increases in the volume of trade between the two countries. Prior to 1957, for example, trade relations between the two countries were negligible. Since then, the US exports to, and imports from, Tunisia have significantly increased.[76]

Insofar as Morocco is concerned, American indecisiveness was evidenced during and after the World War II periods. American–Moroccan relations during the second phase started when the Allied forces, under General Dwight Eisenhower, landed in Morocco in 1942, liberating the country from the French pro-Marshal Pétain Vichy's administration. Instead of granting Morocco its independence, the Allied forces placed the country under the control of General Charles de Gaulle's Free French forces. Official American–Moroccan contacts, however, did not take place until January 1943, when President Franklin D. Roosevelt, British Prime Minister Winston Churchill, and French generals de Gaulle and Giraud met in Casablanca to discuss military operations in North Africa. It was during this occasion that President Roosevelt met with the Moroccan Sultan Mohammed Ben Youssef, 'and no doubt impressed him with the American view that the end of colonial rule would be one of the inevitable consequences of the war'.[77] According to Ronald Steel, the meeting had a positive impact on Sultan Mohammed. From then, onwards, the Sultan increasingly opposed the French admin-istration of Morocco, refusing to countersign laws and rejecting all French compromise offers, holding out for unqualified independence and the abolition of the Protectorate.[78]

The Sultan's alliance with the nationalists and his open demand for independence did not please the French, nor did it incite the US govern-ment to use its leverage to persuade the French government to grant Morocco its independence. As a result of this American aloofness, on 20 August 1953, Sultan Mohammed was arrested by the French Resident-General, General Guillaume, banished with some members of the royal family to Madagascar, and was subsequently dethroned by the French.

Though the incident served to trigger an open Moroccan resistance to French rule throughout the country and in certain respects had adversely affected the Allied war effort, the US government did not budge. Nor did it express any concern regarding the Moroccan insurgency. More import-antly, in 1950 the US reached agreements with the French authorities to establish several air fields and bases on Moroccan territory.[79] The agree-ments were concluded without any consultation with the Moroccan Sultan (later King).[80] In fact, they were concluded in complete disregard to the deposed Sultan and the wishes of the Moroccan people. When the French government realized that it had no choice but to return Sultan Mohammed to his throne in 1955, and to grant Morocco its independ-ence in 1956, both the US and Morocco had to address the issue of the bases and their legitimacy.

The first American–Moroccan crisis in the post-World War II period occurred when '[American] General [Lauris] Norstad invited . . . Sultan [Mohammed] to visit the [bases]. He politely declined; despite their enormous size, Morocco didn't officially acknowledge [the legitimacy of] their existence'.[81] When the US government offered to work out a new lease for the bases with Morocco, and if necessary to renegotiate some clauses contained in the original agreements (which were con-cluded with France), the Moroccan government preferred not to commit itself to agreements that would make Morocco a party to the East–West conflict, or a major target in case of a war between the opponents of the Cold War. In December 1959, both countries reached an agreement that provided for the total withdrawal of all US military forces from Morocco by the end of 1963.[82] The withdrawal of US military forces from Morocco was based on a mutual agreement, a factor that contributed to friendly relations between the two countries thereafter, especially when King Hassan succeeded his father to the throne of Morocco in 1961. This friendship was manifested in frequent visits by King Hassan to the US, a subsequent agreement between the governments of both countries to allow regular US Navy ships to visit Moroccan ports, the US becoming a major supplier of arms to Morocco, King Hassan's opposition to communism, Morocco's support for American peace initiatives in the Middle East in general and the Middle East peace process in particular, and increased American aid to Morocco.

Before Morocco's independence, the US channelled its development grants to the country through the French administration. After Morocco's independence in 1956, American aid (estimated at over $1 billion between 1956 and 1990) went directly to the Moroccan government.

The importance of Morocco to US Mediterranean and global strategic interests was perhaps best exemplified by the fact that prior to 1963, Morocco was a recipient (aside from US expenditures on the air bases) of 'almost half the total sum allocated [by the American government] to all Africa ... '[83] American-Algerian relations during the second phase of US foreign policy began when Allied Anglo-American forces landed at Sidi-Ferruch in November 1942. The landing of these forces had a powerful impact on the Algerian nationalists who apparently were 'aware of the anticolonialist creed of Roosevelt's America'.[84] Ferhat Abbas, an Algerian nationalist leader, seized this opportunity to meet with Bob Murphy, the personal representative of President Roosevelt in Algiers, 'to explore the possibility of applying the Atlantic Charter to Algeria'.[85] The meetings, however, were of no practical consequence. The US was committed to helping Free France and moreover, regarded Algeria as a French territory and hence, a French concern.

Though American–Algerian relations started to assume a new course during the Eisenhower Administration (1952–60), this policy continued virtually throughout the period of Algeria's war of independence between 1954, when the Algerian Front de Libération National (FLN) staged its insurgency, and 1962, when France formally granted Algeria its independence.

The change was first evidenced after Vice-President Richard Nixon's visit to Tunis in March 1957. This visit followed an Arab 'summit' in Cairo which declared the total support of Arab leaders for Algeria's struggle for independence. The Summit expressed a growing awareness of the need to address the grave deterioration in Arab–American and American–North African relations.

The occasion of Nixon's visit culminated in a report in which Nixon predicted that 'the course of [Africa's] development could well prove to be the decisive factor in the conflict between the forces of freedom and international communism'.[86] More importantly, Nixon 'recommended the creation of a separate Bureau of African Affairs in the State Department'.[87] Regarding Algeria, Nixon proposed 'a referendum whereby the Algerians could choose freely between the *loi-cadre* statute being prepared by Robert Lacoste, Algeria's French governor-general, or total independence'.[88] Evidently, the proposal signaled the start of a new American position towards both France and Algeria. For the first time since the Second World War, an American public official pronounced the right of Algerians to self-determination.

Nixon's recommendations to President Eisenhower were adopted, and in 1958, the National Security Council (NSC) delved into a series of policy reviews the aims of which were to halt the spread of Soviet influence on the African continent, as well as to win the friendship of African territories, including those of North Africa. Clearly, the shift in US foreign policy helped facilitate the realization of FLN's goal of achieving Algeria's independence, which it attained on 5 July 1962.

Between 1958 and 1962, the US foreign policy towards the Algerian question focused on avoiding any association with France's colonial policy, pressuring France to grant Algerians the right to determine their future, and developing appropriate grounds for direct peace talks between the FLN and the French government. This policy was deemed least harmful to US interests in North Africa, especially following the establishment of the Provisional Government of the Republic of Algeria (GPRA) with headquarters in Tunisia on 19 September 1958.[89] American aid to Algeria remained limited for decades.[90] In late February 1992, however, further American help to Algeria was part of a general $1.45 billion Western credit package that was extended by 'a consortium of Western banks, under pressure from Western governments', including France and the US.[91] American help was politically motivated.[92] In all probability, its undeclared purpose was to abet the anti-Islamic fundamentalist military leaders of Algeria stabilize their regime, surmount their domestic difficulties that they have been facing since their *coup* in late 1991, and subsequently, led to the cancellation of the first round of elections that were won by the Islamic Salvation Front (FIS) during the same year.

American–North African Relations in the Post-Cold War Era

The end of the Cold War has impacted US foreign policy in more than one way. It has fathered a change in its global strategic interests and priorities, and turned some areas of the world which were strategically important during the Cold War into marginal regions. This generalization, however, does not generally apply to North Africa. North Africa will probably continue to be central to the US for a long time into the future. As long as the US is committed to the promotion of international trade, the defense of Europe, and the protection of Middle East oil resources, developments in North Africa and its constituent states will continue to occupy American foreign policymakers. The region will continue to be vital to US interests because of geopolitical, military, demographic, and economic considerations.

Chief among these considerations is the geostrategic location of North Africa.[93] The region contains not only crucial east–west routes that connect Europe, hinterland Africa, and Asia, but also supports many land, sea and air communication centers and vital seaways such as the Suez Canal, the Strait of Sicily, and the Strait of Gibraltar. Its countries have many important harbors, naval and repair bases and bunkering stations, and more importantly, command and control over shipping lanes in the southern Mediterranean and Red Seas and the outlets of both seas to the Indian and Atlantic Oceans.[94]

The strategic importance of the region is illustrated by the following five features:

1.   Europe and North Africa are separated by only a few miles at the Strait of Gibraltar (maddiq jabal Tariq) in the west. This strait, a channel connecting the Mediterranean Sea with the Atlantic Ocean, is one of the busiest interoceanic waterways in the world, with about 150 ships a day passing through.[95]

Sea access to or from the Mediterranean would require passing through the narrow Strait of Gibraltar, which is under the control of both Morocco and Spain.[96]

As a global power with global trading interests and commitments to the defense of Europe and Middle East oil, the US has a stake in whatever relates to the security of the Strait of Gibraltar, as well as to the freedom of shipping through it. Hence, securing friendly relations with Morocco cannot but continue to be an American concern. By extension, the US is compelled to concern itself also with internal developments in Morocco, lest adverse developments lead to unfriendly regimes.

2.   Europe and North Africa are separated by 65 to 100 nautical miles at the Strait of Sicily, which is located at the ridge that separates the island of Sicily from North Africa, and divides the Mediterranean Sea into eastern and western parts. Tunisia is about 65 nautical miles from Pantelleria Island, a small island lying southwest of Sicily, and about 100 nautical miles from Sicily, the southernmost part of Italy. Hence, access to the eastern or western parts of the Mediterranean Sea, especially at times of war, may depend on the strategic interests of both Italy and Tunisia, if not Algeria, France, and Libya.[97]

The US has a stake in whatever relates to the security of the Strait of Sicily, as well as to the freedom of shipping through this strait's navigable channels. The presence of the US Sixth Fleet in the Mediterranean Sea is to protect not only NATO's southern flank in the Mediterranean Sea but also other diverse American interests in this area.

3.   The continents of Asia and Africa are joined at the Isthmus of Suez, between Africa proper and the Sinai Peninsula.[98] The Isthmus of Suez is linked to the Mediterranean Sea by the Suez Canal (north), which links the Atlantic and Indian Oceans via the Mediterranean and Red Seas. The opening of the Suez Canal in 1869 revolutionized the strategic importance of the North and East African states of Egypt, Sudan, Djibouti, Eritrea, Ethiopia, and Libya, and incited rivalry between France and Britain, to control this crossing point and the countries adjacent to it.[99] A voyage via the Suez Canal may mean saving, on the average, 17 to 18 days over a trip made via the Cape of Good Hope in South Africa.[100]

The Suez Canal is today under the control of Egypt. Reaching it, however, may be influenced by neighboring littoral states, especially Libya, Eritrea, Saudi Arabia, Sudan, and Yemen. Presently, about 60 ships pass through this canal every day.[101]

The role of the Suez Canal has acquired special importance to trading nations of the developed and developing worlds. It has acquired special importance as a main oil transportation route, especially to Western Europe and the United States. Aside from what is delivered through the Suez–Mediterranean pipeline (SUMED), approximately '10 percent of Gulf region oil exports to the US and Western Europe now go through the Canal'.[102] For this reason, it will continue to be a central concern of US foreign policy in the area.

4.   The continents of Asia and Africa are separated by a few miles at the southern end of the Red Sea. This strait, known as Bab el-Mandab, connects the Red Sea with both the Arabian Sea and the Indian Ocean.[103] This channel too is important to the free flow of global trade.[104]

Freedom of shipping through the Isthmus of Suez and the Red Sea are vital to the United States. As long as the US is committed to the defense of Europe and the oil in the Arabian peninsula, the free flow of oil resources to North America, Western Europe, as well as to the promotion of world trade and American access to international shipping routes, Egypt and Sudan will continue to be central to the US post-Cold War global strategic interests.

5.   The coastline of North African countries is 8,201 kilometers long. Of this total, 998 km are located within Algeria, 2,450 km (Red Sea and Mediterranean) within Egypt, 1,770 km within Libya, 1,835 km (including the Atlantic) within Morocco, and 1,148 km within Tunisia. Libya has the longest Mediterranean stretch of this coastline, and moreover, claims the Gulf of Sidra as a Libyan gulf. This feature places navigation in the southern Mediterranean and portions of the Red Sea at the mercy of North African states. Because of these features, North African states and their developments have been, and probably will continue to be, central American concerns. In certain regards, they explain why, as noted by Alastair Drysdale and Gerald H. Blake, 'US military aid is heavily concentrated in states bordering or close to strategic waterways'.[105] Regarding military considerations, North African countries are ideal as access points to troops and military equipment *en route* to Europe in general and southern European countries in particular, the Arabian peninsula, and the Fertile Crescent countries. Their territories are ideal to support NATO's southern flank in the Mediterranean, as well as to the defense of both Europe and the Middle East.

During the Second World War, the British and Allied headquarters in Cairo were named as 'HQ Middle East' and covered large parts of northern and eastern Africa, all the Arab states east of the Suez Canal, in addition to Iran, Turkey and Cyprus. More recently, North Africa, especially Egypt and Morocco, proved to be indispensable bases during the Gulf War. Because of such military geopolitical considerations, North African states and their stability will continue to be considered within the United States' 'high politics.'

Regarding demographic considerations, North Africa will continue to be vital to US interests because its peoples are African, Mediterranean, Islamic, Arab and Berber, and are influenced politically, culturally, and economically by Europe. This suggests that northern African countries are vulnerable not only to whatever developments take place within their respective countries, but also to whatever takes place in this vast and often volatile region. North Africa's Arab/Berber and essentially Muslim character, make its countries part of the Arab and Muslim worlds, and hence, subject to influences emanating from these worlds, as well as a factor in whatever occurs in them. Its location makes its countries part of the African continent, providing its constituent states with ample opportunities to impact and be impacted by sub-Saharan African developments. Its economic and cultural linkages with southern European states expose its countries to the strains and stresses of European politics. By the same token, developments within its countries affect European countries and their respective interests. For all these reasons, the politics of the states of North Africa are interconnected with those of other regions of the world. Jointly, they support a regional balance of power system with its own actors, its own rules, and its own values. At times, this power system operates independently from other systems. At other times, it converges, diverges, or overlaps with those of Africa, the Arab and Muslim worlds, and Europe, if not the international system as a whole. Hence, no country can enter into relations with a North African state without regard to the wider region, as well as to the international interests of its constituent countries. This explains why the US has emphasized the regional dimension of the Middle East peace process, involving several North African states in its talks, and proposing regional economic integrational projects such as the Middle East and North Africa Development Bank. Despite reservations by some European states (including France and the UK), the bank was established with headquarters in Egypt.

Finally, North Africa will continue to be vital to US interests because of economic considerations. Economically, North African countries support open markets, possess resources, and have facilities that are vital to the well-being of Western Europe and the United States. In addition, their location allows them to provide facilities for intercontinental transit trade operations between North America, Europe, hinterland Africa, and western Asia. Furthermore, their markets are in dire need of Western manufactured goods, food products, and technology, which the United States can provide. Until the Second World War, the economies of North African countries and those of western Europe were heavily linked, if not integrated, with each other. Since the Second World War, and since the independence of North African countries, the US, slowly and gradually, has been successfully penetrating North

African markets and expanding its economic interests in them. This was done, however, without disrupting the region's trading partnerships with European countries. Libya however, is presently the only North African state with which the US is not a major trade partner.[106]

Evidently, these growing relations between North Africa and the US were realized despite North Africa's traditionally close economic and financial linkages with European countries. To all indications, the future will show greater competition between American and European interests in North African markets. Already, the European Communities (EC) and member states have entered into several cooperational agreements with the Arab Maghreb Union (AMU). This latter body was established in 1989, to promote cooperation and integration among its five North African members (Algeria, Libya, Mauritania, Morocco, and Tunisia), and recently was joined by Egypt as an observer.

Jointly, the countries of North Africa support a market of about 163 million people. This market is rich in natural resources that are vital to the well-being of Western Europe and industrialized countries elsewhere. These resources include oil, natural gas, and numerous diverse mineral deposits. North Africa's proven oil reserves range from 38.2 to 53.6 billion barrels, nearly 4 to 5.5 percent of total estimated world reserves. Of all its countries, Libya holds the largest share (22.8 to 38.4 billion barrels). Most of North Africa's oil exports are directed to Europe.

Moreover, North African countries account for 174.9 to 186.8 trillion cubic feet of the world's (4,375.8 to 4,675.9 trillion cubic feet) natural gas reserves. Of all North African countries, Algeria holds the largest share of 116.5 to 128.1 trillion cubic feet – a share that also makes its known reserves the fourth largest in the world. Most of North Africa's natural gas exports are also directed to Europe. Furthermore, North Africa is rich in numerous mineral deposits, especially phosphate. Morocco alone 'possesses more than 60% of the world's proven phosphate deposits and is the world's largest phosphate exporter'.[107]

Regarding transport facilities, most countries of North Africa support seaports and relatively modern airports. The markets, natural resources, and facilities of North African countries have attracted US private investment, since the 1950s. Most of these firms specialize in petroleum and mining or related activities, and some of them, as in Algeria, have joint venture agreements with state enterprises. In addition, many American trading firms have been exporting their products to several North African countries, through the use of Export–Import Bank facilities.

In light of these considerations, it is quite evident that North Africa is economically advantageous to the well-being not only of Western Europe, but also of the United States. Even if US economic interests are disregarded, North Africa cannot but be a vital concern to the US in the post-Cold War era. American commitment to the defense of Europe cannot be attained without also protecting European economic interests, especially in those areas of the world where these interests are deeply ingrained, as they are in North Africa.

United States' policy towards North Africa in the post-Cold War era appears to be one that emanates from numerous factors, including the United States' commitment to the defense of Europe, the protection of the access to oil from the Arabian peninsula, safeguarding the shipment of oil to the West, and freedom of trade, shipping, and communications. North Africa's geopolitics are directly linked to these objectives, which are currently illustrated in American attempts to cultivate friendly relations with the countries of North Africa, secure the stability of the region, support North African friendly governments, involve North African countries in the Middle East peace process, and create conditions and deploy means conducive to the moderation of religious extremist movements should they threaten regional security.

By far, the latter threat is most critical. This threat poses itself not only as a serious challenge to the US interests in the area, but also to those of several allies in the region. More importantly, it impresses upon the US the need to find a means of helping its allies in North Africa to narrow the factors underlying the spread of religious extremist movements, if not neutralize their anti-Western orientation.

So far, however, religious extremism in North Africa as elsewhere in the Arab/Muslim Worlds continues to be a highly misunderstood phenomenon in the US. Clearly, as long as this misunderstanding remains, and as long as the political and economic factors underlying the rationale of religious extremism persists, American foreign policy objectives in North Africa will continue to be under siege in some quarters.

Notes

1. James Schlesinger (January 1993) 'Quest for a Post-Cold War Foreign Policy', *Foreign Affairs*, 72 (1), 17–18.
2. Ibid., 18.
3. See Marguerite Michaels (January 1993) 'Retreat From Africa', *Foreign Affairs*, 72 (01), 93–108.
4. See Michael Clough (May 1992) 'The United States and Africa: The Policy of Cynical Disengagement', *Current History*, 91 (565), 193–8.

5. See Thomas M. Callaghy (January 1994) 'Africa: Falling off the Map', *Current Affairs*, 93 (579), 1–36; and Michael Chege (January 1994) 'Remembering Africa', *Foreign Affairs*, 71 (1), 145–63.
6. Philip W. Bell 'Colonialism in American Foreign Policy', in Harold Karan Jacobson (ed.) (1960) *America's Foreign Policy* (New York: Random House), 507. For further treatment of this phase, see Michael Hunt, 'Traditions of American Diplomacy: From Colony to Great Power', in Gordon Martel (ed.) (1994) *American Foreign Relations Reconsidered 1990–1993* (New York: Routledge), 2.
7. For further details, see John A. Garraty and Peter Gay (eds.) (1981) *The Columbia History of the World* (New York: Harper & Row), 760.
8. Harold D. Nelson (ed.) (1978) *Morocco: A Country Study* (Washington, DC: The American University Press), 40.
9. Ibid.
10. Ibid.
11. Ibid.
12. Tangier, a strategic port, which had been under international administration (with foreign relations reserved to the French) since 1923, was ceded by the Spanish government to Morocco in 1969.
13. Algeria was nominally governed by Ottoman rulers between 1529 and 1830. For privileges covered by the capitulations agreement, see Lord Kinross (1977) *The Ottoman Centuries, The Rise and Fall of the Ottoman Empire* (New York: Morrow Quill), 204; William R. Polk (1991) *The Arab World Today* (Cambridge, MA: Harvard University Press), 84; and *The New Encyclopaedia Britannica* (15th edn, 1988), 2, 832.
14. Nelson D. Harold (ed.) (1986) *Algeria, A Country Study* (Washington, DC: The American University, 27–8, and Lord Kinross (1977) *The Ottoman Centuries, The Rise and Fall of the Ottoman Empire* (New York: Morrow Quill), 218.
15. Harold (1986), 27–8.
16. Ibid., 28.
17. Jane Soames Nickerson (1961) *A Short History of North Africa From Pre-Roman Times to the Present* (New York: The Devin-Adair Co), 96.
18. Talcott W. Seelye (March 1990) 'Ben Ali Visit Marks Third Stage in 200-Year-Old US–Tunisian Special Relationship', *The Washington Report on Middle East Affairs*, 8(11), 7.
19. Ibid.
20. Ibid.
21. Ibid.
22. Ibid.
23. Ibid.
24. The Pasha was from the Karamanli family of (western) Tripoli, Tarabulus al-Gharb. The Karamanlis ruled parts of what is presently modern Libya from 1710–1835 under nominal Ottoman sovereignty.
25. Polk (1991), 156.
26. This exploit earned Decatur 'his captain's commission and a sword of [honor] from Congress.' *The New Encyclopaedia Britannica* (15th edn, 1988), 3, 948.
27. Nickerson (1961), 95.

28. The peace treaty also provided for the exchange of prisoners, as well as for a financial settlement amounting to $60,000 payable by the US, to close the contract. Ibid., 95–6.

29. Ibid., 96.

30. Harry N. Howard, 'The United States and the Middle East', in Tareq Ismael (ed.) (1974) *The Middle East in World Politics* (Syracuse: Syracuse University Press), 117. Quoted by Alan R. Taylor, 'The United States and the Middle East', in Tareq Y. Ismael (1986) *International Relations of the Contemporary Middle East* (Syracuse: Syracuse University Press), 134.

31. For details regarding the mission see: Clarence Clendenen, Robert Collins, and Peter Duignan (1966) *Americans in Africa 1865–1900* (Stanford: The Hoover Institution), 75–6.

32. Upon driving Egyptian forces from Syria and Arabia, the British and Ottoman governments forced Mehmet Ali to reduce his forces (from over 130,000 to 18,000) and to abandon his industrialization and reform programs.

33. For a review of the rise and de-empowerment of Mehmet Ali, see Polk (1991), 79.

34. Ibid., 81.

35. Ibid., 384.

36. Ibid.

37. Garraty and Gay (1981), 1006.

38. Ibid.

39. In 1943, for example, Hooker Doolite, American Consul to Tunis, tried 'to foil the insidious attempts of domineering French intent on getting rid of [Habib] Bourguiba, then a rising Tunisian nationalist leader'. According to the Tunisian Secretariat of State for Information, Doolite gave Bourguiba 'full and active support'. Incidents of this nature, however, appear to have been rare, and it is highly likely that Doolite's support was personal rather than in pursuance of American foreign policy directives. Regarding this issue, see Secretariat of State for Information (ed.) (1976) *Tunisia Moves Ahead* (Tunis, Tunisia: Ceres Productions), 44.

40. Polk (1991), 385.

41. Lawrence G. Potter (1994) 'Islam and Politics: Egypt, Algeria and Tunisia', *Great Decisions* (New York: Foreign Policy Association), 75.

42. Polk (1991), 159.

43. For further analysis, see: Michael Clough (1992) *Free at Last? US Policy Toward Africa and the End of the Cold War* (New York: Council on Foreign Relations Press), 5.

44. Alan R. Taylor 'The United States and the Middle East', in Ismael (1986), 140.

45. The United States' major preoccupation at the outset of the Cold War was to 'employ any means at its disposal to prevent the Soviet Union from filling the power vacuum ... created by the incipient withdrawal of the old colonial powers'. Ibid.

46. Ibid., 144.

47. For further analysis, see Alastair Drysdale and Gerald H. Blake (1985) *The Middle East and North Africa* (New York: Oxford University Press), 307–8.

48. Denis J. Sullivan (October 1996) 'American Aid to Egypt, 1975–1996: Peace Without Development', *Middle East Policy*, 4(4), 36. For further details see Drysdale and Blake (1985), 307–8.

49. Sullivan (1996), 37.
50. Helen C. Metz (ed.) (1991) *Egypt: A Country Study* (Washington DC: The American University Press), 331–2.
51. Ibid., 314, 332.
52. Ibid., 332.
53. Drysdale and Blake (1985), 34.
54. Colbert C. Held (1994) *The Middle East Patterns, Places, Peoples, and Politics* (Boulder: Westview Press), 374.
55. Sinclair Road (December 1996) 'Britain's Trade with the Middle East and North Africa', *Middle East International*. 540, 21.
56. Polk (1991), 160. Because of strategic considerations, the Soviet Union demanded the right to control both Libya and Massawa, Eritrea. This demand was calculated as a means to establish control over the eastern Mediterranean and the Red Sea, and practically, over the Suez Canal. See Olatunde J. C. B. Ojo, D. K. Orwa, and C. M. B. Utete (1985) *African International Relations* (London: Longman), 96.
57. Polk (1991), 160–1.
58. Skyne Uku-Wertimer (1990) *Africa Changes and Challenges* (Acton: Tapestry Press), 24.
59. Wayne C. McWilliams, and Harry Piotrowski (1993) *The World Since 1945* (Boulder: Lynne Rienner), 397.
60. According to McWilliams and Piotrowski, President Reagan provided no proof of Qadhafi's complicity, and, moreover, West German intelligence stated that the evidence pointed to Syria. Ibid., 397, 408.
61. Ibid., 398.
62. George Joffe (6 December 1996) 'Libya Turning the Tide', *Middle East International*. 539, 14.
63. The American Assembly (June 1958) *The United States and Africa, final edition of background papers prepared for the use of participants, and the final report of the Thirteenth American Assembly* (New York: Columbia University Press), 23.
64. Ibid.
65. Ibid.
66. Quoted by Clough (1992), 5.
67. Alistair Horne (1987) *A Savage War of Peace, Algeria 1954–1962* (New York: Penguin Books), 15–16, 249–50.
68. Ibid., 475.
69. Ibid., 15–16.
70. Ibid., 243.
71. Harold D. Nelson (ed.) (1986) *Tunisia: A Country Study* (Washington, DC: The American University Press), 264.
72. American aid to the Tunisian government was sustained in the following years with continued American pronouncements expressing the United States' firm commitment 'to the sanctity of Tunisia's territorial integrity and to the principle of noninterference in its internal affairs'. Ibid., 264.
73. Ibid., 306.
74. Ibid., 306–7.
75. Ibid., 264.
76. In 1995, for example, the US exports to Tunisia accounted for 5 percent of Tunisia's total imports ($6.4 billion), whereas US imports from Tunisia

accounted for 1 percent of the country's total exports ($4.1 billion). Central Intelligence Agency (1995) *The World Factbook, 1995–1996* (Washington, DC: Brassey's), 400.

77. Ronald Steel (1967) *North Africa* (New York: The H. W. Wilson Co), 61.
78. Ibid.
79. Ibid.
80. Nickerson (1961), 182–3.
81. Steel (1967), 92.
82. American airbases were vacated in December 1963. The Kenitra naval facility, however, was not closed until 1978.
83. Nickerson, 188. Also (1995) 'Foreign Commerce and Aid', *Statistical Abstract of the United States, 1995* (Washington, DC: The Reference Press), 810. For further details see also the annual reports of the US Agency for International Development, *US Overseas Loans and Grants and Assistance from International Operations* and the US Department of Commerce report (1995) *US Direct Investment Abroad* (Washington, DC: US GPO).
84. Horne (1987), 42.
85. Ibid.
86. Quoted by Clough (1992), 6.
87. Ibid.
88. Horne (1987), 247.
89. Ibid., 404.
90. See 'Foreign Commerce and Aid' (1995), 810.
91. Howard LaFranchi (3 March 1992) 'Algeria Receives Western Aid to Stabilize Economy', in *The Christian Science Monitor*. 6.
92. During the same year, Algeria received $1 billion in loans from the European Community, the World Bank, and the International Monetary Fund (IMF). Ibid.
93. North Africa is a region where Eurasian and African continental land masses converge, a feature that comprises one of the world's great human and geographical junctions.
94. Insofar as the defense of Europe is concerned, World War II has made evident the strategic importance of North Africa. Short of nuclear war and intercontinental ballistic wars, the continued strategic significance of North Africa to the defense of Europe cannot be disclaimed. As regards free trade and freedom of navigation in the Mediterranean and Red Sea, past and recent crises confirm that they cannot be achieved without the full cooperation of North African states. They also suggest that threats to these objectives are likely to recur in the future, more so, because of bilateral, regional, and international crises, as well as the different practices of states with regard to the United Nations Convention on Law of the Sea of 1958 in general, and the right of 'innocent passage' through the territorial waters of littoral states. Some states, such as Algeria, 'require foreign warships to obtain permission before entering their territorial seas'. Drysdale and Blake (1985), 13.
95. The Strait of Gibraltar lies between Spain, in Europe, and Morocco, in Africa. It is 35 nautical miles long and its narrowest points between are only 8–10 nautical miles wide. Its depth varies from 270 to 2000 feet. For further details, see ibid. 136–7.

96. During the Punic Wars, the Gibraltar Strait was crossed by Hannibal, the commander of the Carthaginian Empire in what is presently modern Tunisia, when he launched his campaign against Rome. In the seventh century it was crossed by Tariq bin Ziyyad, a Muslim commander, in a campaign against the Iberian Peninsula and southern France. In the 1930s, it was used by General Francisco Franco to stage his revolution against the government of Spain.
97. The navigable channels of the Strait of Sicily that are regarded international waters narrow down significantly to about 44 nautical miles, more so, when the territorial waters and maritime claims of the coastal states are taken into consideration.
98. The Suez Canal is 110 miles long. Since its excavation between 1859 and 1869, it has been enlarged at least nine times, the last in 1980. In early 1990s, Egypt started a new expansion project to allow for the passage of large tankers. As a result of this project, the Canal's navigational width and depth will increase from 590 feet to 1,360 feet and from 52 feet to 72 feet, respectively. Held (1994), 195.
99. Ibid.
100. Using the Suez canal may cut the distance up to half. For example, the distance between New York to Bombay, India, via the Cape of Good Hope is 11,800 nautical miles; whereas it is 8,200 nautical miles via the Suez Canal. Using the Suez Canal means shortening the distance by 31 percent. In another example, the distance between New York and Ras Tanura, Saudi Arabia, is 11,900 nautical miles by the Cape of Good Hope, whereas it is 8,300 nautical miles by the Canal. This cuts the distance by 30 percent. The implications of these savings on overhead costs as well as on fuel costs are self-evident. For further details, see Drysdale and Blake (1985), 13, 131.
101. Ibid., 13.
102. Ibid., 141, 143.
103. It is 35 nautical miles in length, 10.5 nautical miles in width, and 40–600 feet in depth. For further details, see ibid., 141, and Held (1994), 195, 372.
104. Drysdale and Blake (1985), 130.
105. Ibid.
106. The US now ranks third among Algeria's major export partners, and fourth among its major import sources; second among Egypt's major export and import partners; fourth among Morocco's major export partners and second among its import partners; and fifth among Tunisia's major export partners and second among its major import partners. Trading relations between Libya and the US are presently strained and have been so since Qadhafi's *coup* of 1969. Today, the US ranks third among other exporters to North Africa. In 1995, its exports to the region amounted to $378.3 million per month, ranking third after France and Italy. For further details, see *The World Factbook, 1995–1996*; and Sinclair Road (20 December 1996) 'Britain's Trade with the Middle East and North Africa', *Middle East International*. 540, 20–1.
107. Robert Famighetti (ed.) (1994) *The World Almanac and Book of Facts, 1994* (Mahwah, NJ: World Almanac), 155.
108. Uku-Wertimer (1990), 32.

2
Post-Cold War Era West Africa: Implications for US Foreign Policy

Earl Conteh-Morgan

Introduction

Africa, perhaps more than other regions of the world, has always been profoundly subject to the ebb and flow of global systemic values and changes: colonization and decolonization, superpower ideological rivalry, the demise of the Cold War, and authoritarian tendencies and democratization, among others. As waves of change roll over the international environment, so do the policy postures of great powers also change vis-à-vis the various nations of the African continent. The history of Africa's relationship with the great powers has, accordingly, been one where Africa has assumed varied episodic geopolitical utilities shaped by the rhythm of global geostrategic and military rivalries.

As Africa continues to be challenged by this new *fin de siècle* international environment, it is confronting some formidable sociopolitical and economic problems unprecedented in post-World War II era international politics. In particular, West Africa, is undergoing dramatic change with profound politico-economic implications for the heightened interdependence of the international system. Several self-declared Marxist regimes have modified their ideological inclinations with the demise of the Soviet Union. Pro-free market regimes (Ghana, Sierra Leone, Togo, and Nigeria, among others) have embarked on more privatization as a way of decreasing state control over their economies. The decade of the 1990s which produced a new world political environment also produced new developments in the West African sub-region: a fragile security situation in Liberia, where the attempted power-sharing formula among warring factions has remained unstable; the grip of anarchy in most of Sierra Leone as a result of the spillover of the Liberian conflict in 1991; a deteriorating political situation in Guinea which in 1996 experienced a failed *coup d'état*; the likelihood of Senegal's Casamance province insurgency spreading to a wider part of the country; a seemingly explosive and potentially dangerous situation in Nigeria intensified by repression and a tenacious resistance to democratic change by an entrenched and self-propagating military oligarchy; and the pressure of massive refugee movements within the sub-region due to conflict situations and increasingly fragile economies.[1] The entire sub-region continues to be weighed down by the economic austerity of Structural Adjustment Programs (SAPs), and the perennial lack of foreign exchange to service the ever mounting burdensome debts. Directly related to the above are the pressing domestic realities of collapsing and deteriorating domestic socio-economic sectors such as public health, education, agriculture, and ecological institutions.

All of the above developments mean that the end of the Cold War acts as an imperative for American policy to view Africa from a new perspective. In particular, since West Africa is also in a state of change and continuity, America's relations with many countries in the region will be played out in a much different policy environment than in the past. The fundamental issue is how post-Cold War era American foreign policy addresses the West African sub-region. In other words, what is the current pattern or trend in American foreign policy to the region? And what are the parameters of change and continuity in US West African policy since the end of the Cold War?

West Africa's human needs are not isolated from the world's political and economic problems. Since the end of the Cold War, the principles underlying Presidents Bush's and Clinton's statements on Africa point to the importance of support for democratization and human rights, reform of development assistance, and strengthening of UN peacekeeping efforts, among other concerns. In addition, the US now has an urgent interest in ameliorating a set of common global concerns – poverty, environment, population growth, AIDS, and drug addiction – that in different ways threaten the well-being of Americans domestically. Accordingly, this chapter examines the extent to which the US is pursuing the above objectives by (1) defining America's broad interests in West Africa; (2) examining what is/or has been US policy since the end of the Cold War; and, (3) if necessary, exploring what directions US foreign policy ought to take towards the sub-region.

Africa's Geostrategic Importance: Myth or Reality?

It has become a trend in United States foreign policy to reconfigure its interests, concerns, and anxieties regarding Africa. In other words,

Africa ebbs and flows into US geostrategic preoccupations, sometimes of immense instrumental value, at other times as a continent to be treated with benign neglect.[2] Moreover, since its initial contact with the West, Africa has been the subject of varied images and differing guises depending on the prevailing tempo and dynamics of global power politics.

As far back as the early years of European exploration in the sixteenth century, Africa's geographic location made it an obstacle to the access to more important regions of the Middle East or Asia. But from the perspective of Middle Eastern and Far Eastern regions, Africa has been of military-strategic value because of its utility as a buffer, or protection from Western interventions. Although initially an obstacle to European explorations, subsequently locations along the African coast became refreshment stations or way stations on the perilous and long commercial voyages to the Far East. In addition, Africa's role as a means to an end (the objective of arriving at the Far East) increased when way stations built by competing Europeans (Portuguese, Dutch, French, and English) became fortified. They served as protective locations and as bases from which to attack enemy ships. Accordingly, Africa's choke points at the eastern and western ends of the Mediterranean and the Cape became objects of rivalry and control. It also meant the control of geostrategic enclaves to protect access to lucrative markets in the East, as well as to safeguard seaborne commerce. Furthermore, it led to the rivalry, as early as the 1650s, between Spain and Portugal, the Netherlands, and later between Britain and France. The scramble for Africa in the 1880s, received its primary motivations in European perceptions of geostrategic imperatives.

In times of war way stations or refreshment locations became essential points of attack against the enemy. Thus, Africa's military-strategic salience often increased in times of war as a launching pad for defensive and/or offensive purposes. In their geopolitical theorizing, Halford Mackinder and Nicholas Spykman had viewed parts of Africa as essential to the control of the heartland (Germany and/or Russia).[3] Subsequently, with the emergence of Cold War rivalry, Africa became a point of focus in the US strategy of containing the Soviet Union.

Along with the strategic (choke points) preoccupation went the objective of exploiting Africa's critical minerals. The latter goal which is recent, focused mainly on Africa's minerals in relation to their importance in the nuclear and space industries.[4] Uranium from the then Belgian Congo contributed to the success of the Manhattan Project. Since then the importance of Africa's minerals has increased due to the imperatives of rapid technological development.

The rapid retrenchment from the continent by the US and the Soviet Union following the end of the Cold War, and the virtual underdevelopment of state institutions and infrastructure during and following the end of colonial rule, shows that great powers used Africa to fight their geopolitical battles but abandoned it with even more problems when world political conditions transformed. In other words, Africa represents another terrain, receptive, easy to penetrate and manipulate, and ever ready to be used by external interests when it is expedient to do so.

The United States and Africa

Prior to 1945, Africa was viewed by the US as a mere territorial outpost of Europe because of the latter's role as colonial master, and Europe's economic interest in the continent. Accordingly, in the area of policy towards Africa, the US largely deferred to the Europeans. Even at the eve of independence the US registered its support for continuing European colonization, it was led to do so because of the stated concern of 'premature independence', and the United States' ignorance of changing world political attitudes and winds of change that were sweeping across Africa and other colonized territories.[5] Hardly any significant US efforts had been put into formulating new policies to match the effect of global systemic changes on Africa's political economy.

When profound international developments occur, or when US administrations change, the general expectation is usually that US policy toward Africa will be marked by significant change. Presently many observers of US policy to Africa assert that Africa is in danger of being marginalized as 'the Fourth World', an unnecessary burden to the industrialized nations because of its seemingly insoluble problems, and above all, irrelevance to the global economy.[6] In the wake of the Cold War's demise, critics claim, only rare instances of US attention have been focused on the crisis in Liberia, Sierra Leone, and the significant issue of governance in Nigeria. These occasional instances of US policy attention underscore the fleeting importance accorded to Africa in US foreign policy. The common assumption is that US relations with Africa have always been relegated to a lesser level in overall US foreign policy interests, after Europe, Asia, Middle East, and the Americas.

Following the end of the Cold War, what were once considered vital US 'interests' in Africa such as strategic minerals or sea lanes, are now disposed of as myths, or inventions of the Cold War. According to Ambassador Daniel H. Simpson, these myths were motivated largely by the utility of the apartheid regime in South Africa to US global strategy vis-à-vis the Soviet Union, or by the utility of the geostrategic location of Zaire.[7] The significance of the 'Cape sea route' became one of these interests-cum-myths. The fear was that alienating the white minority government in South Africa could result in its overthrow by a pro-Soviet majority-based government that could then allow the Soviets to interdict shipping at the intersection of the Atlantic and Indian Oceans.[8] The end of the Cold War has, no doubt, solidly refocused US foreign-policy thinking away from anxiety over such shipping interdiction and other myths.

Just as the strategic lines of communication (SLOCs) argument has lost much of its credibility, so also has that of strategic minerals. It has now become a cliché to say that Africa abounds in strategic minerals. Among many others are chromium, platinum, cobalt, copper, and oil. The reality is that these constitute more of a commercial interest than a 'strategic' interest, especially now that the Cold War is over. The increasing conceptualization of security in economic terms makes Africa's minerals more of a commercial opportunity than a strategic interest. The Cold War geostrategic minerals argument of the 'resource-denial' intent by the Soviets, or possible control of mineral-rich countries by pro-Soviet regimes, have now become outdated. But even during the Cold War, these arguments had serious weaknesses as evidenced by the strange cooperation between the Cuban military and American oil interests in Angola.

Africa is thrust into a new role with every dramatic change in the tempo and tune of world politics. The recent change in the distribution of power among the great powers has spawned a deemphasis on military-strategic preoccupations. The consequence for Africa has been a military and security retrenchment by the US and Russia; and profound changes in the state of domestic politics in the continent. This change is manifested in the clamor for democratization, and inclination towards privatization of economic activities.

America's Cold War Era Policy Posture in West Africa

The official language that has underlined American policy towards the West African sub-region has consistently emphasized historical ties and outlook between the peoples of the two regions. Links with Liberia's history dominates, a stress of the American role in Liberia's founding in the 1820s as a haven for freed American slaves, and official recognition in 1862 as a full-fledged nation. Accordingly, Washington has always regarded Liberia as an American preserve. In the early decades of the twentieth century, US officials occasionally intervened to put the country's finances in order and pressured the government in Monrovia to make concessions to US businesses, especially the Firestone Rubber Company. During World War II, the US military used Liberia as a transit point for operations in North Africa and southern Europe. After the war the US established several critical communications facilities in the country, and secured special landing rights at Robertsfield airport near Monrovia for US military aircraft.

As a sub-region still under European colonial hegemony, West Africa was not accorded any US foreign policy attention until the late 1950s, with the emergence of the independence era in sub-Saharan Africa. After attending Ghana's independence ceremony in 1957, Vice-President Richard Nixon predicted that Africa 'could well prove to be the decisive factor in conflict between the forces of freedom and international communism'.[9] Accordingly, in September 1958, the US State Department established a new Bureau of African Affairs, under the direction of its own Assistant Secretary of State. The following year, 1959, the Senate and House of Representatives established sub-committees on African issues. Before this period, US policy towards West Africa was based on the assurance, and predicated on the belief, that communism posed no threat to Western interests in the sub-region.

A largely nonchalant American attitude towards West African countries persisted because of a combination of two factors: the exigencies of French African policy and growing strains in Franco-American relations. French officials in the many francophone West African territories (Mauritania, Mali, Senegal, Guinea, Ivory Coast, Togo, Benin, Burkina Faso, and Niger) frequently discouraged what they perceived as American cultural or economic encroachments. In a referendum in 1958, France had made it clear to its colonies that it would permit no middle ground between full cooperation within a French community or total independence with no ties to France. The American attitude towards most francophone African leaders was, therefore, one of caution and devoid of any significant or wide-ranging foreign policy innovations. In other words, West Africa had become an extension of the strong geopolitical orientation of French policy in Africa since 1960. However, this orientation was probably cemented towards a Gaullist antipathy towards 'Anglo-Saxon' issues.[10] Thus, Guinea's breakaway from France's political hegemony did not elicit any immediate American recognition. The US probably did not want to become a thorn in France's flesh. Besides, it was also difficult to identify informed American officials with expertise on the new situation in much of Africa.

However, US hesitation and inaction towards developments in Africa soon changed with the administration of President John F. Kennedy beginning in early 1961. The new president had been the first Chairman of the US Senate Foreign Relations Committee's Sub-Committee on Africa. His first appointment was the Assistant Secretary of State for Africa, former Governor of Michigan, G. Mennen Williams, a man well known for his support of African aspirations. To meet the enormous need for educational and technical training, the new administration launched the Peace Corps, for which West Africa became its second major focus (the first was the Philippines).[11] The Kennedy administration quickly infused a positive spirit into US–West African relations. Kennedy was instrumental in softening the anger of new nationalists like Kwame Nkrumah of Ghana, and Sekou Touré of Guinea.

While President Kennedy and Secretary Williams tried hard not to inject any Cold War attitudes in relations with Africa, there were some within the administration whose outlook saw world politics in such hostile terms. Besides, many at the top of the administration also became obstacles to creating a new emphasis on African affairs by their refusal to elevate Africa to any level of importance in American foreign policy. However, the African political climate of the early 1960s soon made it possible for American policy to view Africa in Cold War terms. Divisions among African states into a Casablanca Group, a Monrovia Group, and a Brazzaville (Francophone) Group, suggested that the official non-aligned posture of African states was not necessarily 'non-aligned'.[12] The West African states of Ghana, Guinea, and Mali belonged to the radical Casablanca Group. As a consequence American policy viewed states like Liberia, Nigeria, and Sierra Leone (members of the Monrovia Group) as moderate. Ghana, Guinea, and Mali became the source of radical nationalist rhetoric. The Soviet Union's warm relationship with these latter countries then reinforced American policy posture to view them in Cold War terms.

The US was probably operating on the belief that it was 'losing' West Africa's 'radical' states (Ghana, Guinea, and Mali,) in a zero-sum game with the Soviet Union. President Kennedy's ambassador to Guinea, for example, made strong efforts to convince its President Sekou Touré that America's new administration was predicated on improved relations with African countries. William Atwood, accordingly lobbied vigorously in Washington for a substantial aid package to develop Guinean industry. The subsequent aid package, while not large, was instrumental in diluting the Soviet–Guinean relationship. Guinea in the end was earmarked for a $25 million aid package that would help secure a small dam and power station, six small plants for processing consumer goods, a turbine and generator, and a vocational training program for Guinean workers.[13]

While the US may have also genuinely sought expanded relations with West African states regardless of the contest with communism, American policy was also guided by the fear that in Ghana, Guinea, and Mali, the process of decolonization was yielding new African governments hostile to Western interests, and therefore inclined to support many aspects of Soviet policy in the sub-region. The US thus allocated increased aid in order to invest in the future development of West African states. Judging from the spirit of the times, it felt impelled to do so mainly because of the 'Soviet threat', viewed as a comprehensive and ambitious program to undermine the independence of these new African nations. While most American leaders may not have been convinced that these radical West African countries were moving inexorably towards the Soviet camp, however, they were nonetheless convinced that such young nations could become victims of Soviet intrigue. Before 1961, the US viewed radical nationalist West African states from a faulty lens. For example, Ghana and Guinea were judged too close to communist ideology and were therefore denied aid.[14] In the case of Ghana, aid that had previously been promised in connection with the Volta hydroelectric project, encountered serious reluctance in terms of implementation. In this sense, America's expected increase in Soviet influence in certain West African states shaped American policy towards the sub-region.

In later years, American policymakers raised questions about Kwame Nkrumah's radical leftist-sounding, pan-Africanist posture and Nkrumah's internal subversion of Ghana's opposition. The concern of American officials was particularly based on Nkrumah's strategy of preventive detention through which he was eliminating political opposition to his policies. Such problems of democracy became so widespread among other states of West Africa as well, that negative external perceptions of the sub-region heightened. Ghana, the first sub-Saharan nation to achieve independence after World War II, degenerated into a failed democracy, while Nigeria, initially referred to as a potential economic giant, became by the mid-1980s an example of the worst form of economic mismanagement. American administrations determined their African policy goals primarily by geopolitical considerations and in reaction to Soviet activities in regions considered as geostrategic to American global interests. This geopolitical orientation meant that African economic development and socio-political problems did not become the concern of American official interests until most colonies attained independence. Except for Liberia, development in the rest of West Africa was considered the responsibility of European, ex-colonial masters. There was passing thought that some US Marshall Plan-type aid would trickle down to the African colonies, thereby freeing the US of any substantial obligation to the sub-region. Liberia, in West Africa, and Ethiopia, in East Africa, received special American economic attention because no Europeans had commitments to their economic develop-ment. But with the rest of West Africa, the American view was that Britain and France had direct responsibility for their former colonies where economic development was concerned.

Defining America's Post-Cold War Era Interests in West Africa

Similar to the foreign policies of many other powerful nations, US policy towards West Africa is also at a crossroads. In general, the difficulty of identifying a coherent and substantive US policy is in part due to the changed geostrategic situation in world politics, which is also directly related to the magnitude of changes in the sub-region of West Africa and within its specific nation-states.[15] Liberia erupted into ethnopolitical violence in 1989, and was still plagued by sporadic outbursts of factional fighting seven years later. In 1991, the Liberian conflict spilled into Sierra Leone generating a total breakdown of law and order in most of that country, and generated a huge outflow of refugees beset by hunger and starvation. The countries of West Africa are currently in the grip of a disturbing trend of military interventions and civil strife. In January 1996, Colonel Barre Ibrahim toppled President Mahamane Ousmane, the first democratically elected head of State in Niger. In February 1996 as well, Guinea came close to a military take over. The Sierra Leone military had just ousted its head of state a week earlier. And Nigeria, the supposed hegemonic state in the sub-region, was in the grip of a self-perpetuating military oligarchy that openly flouted democratic change.

This combined impact of external global systemic imperatives and significant domestic developments has resulted in what seems like a policy vacuum in US relations with West Africa. Generality and lack of country-specific focus, especially in relation to war-ravaged and refugee-burdened countries, is the outcome. This outcome is furthered by US administrations focusing on its own mounting domestic social and economic problems. The preoccupations of the Clinton Administration with general Africa policy have not been substantially different from those of the former Bush Administration. Officially, the emphasis on democracy, human rights, conflict resolution (peacekeeping), and economic reforms have been the overarching themes of US post-Cold War Africa policy. The continuity between the two Administrations is reflected in *National Security Review (NSR) 30* of the Bush Administration. This document prepared on the eve of President Bush's departure from office was designed to shape African policy for most of the 1990s. Another document titled *American Policy Toward Africa in the 1990s* has as its first policy identification the following: 'US interests include promoting peaceful change and conflict resolution, democracy and improved governance, sustainable development, and effective African action on transnational issues, such as AIDS, population growth, and terrorism.'[16] The document, all along underscores the new opportunities and constraints for US Africa policy posed by the changed strategic situation, as well as by African socio-economic realities. The opening paragraph captures the main thrust of the document:

> The dramatic changes underway in post-Cold War Africa present unprecedented opportunities and obstacles. Africans are seeking economic progress and democracy. Beset by ethnic tensions, economic decline, and new threats from AIDS or narcotics, Africans increasingly look to the US for assistance and mature partnership while taking more responsibility for solving their own conflicts and problems. The US can best support reform and avoid the costs of instability and humanitarian relief through an agenda of active diplomacy.[17]

The main issue becomes how theory as contained in a document, has conformed to, or been different from, practice. In other words, how effective has been US post-Cold War West African policy as reflected in the broader guidelines of *NSR 30* and those of the Clinton Administration?

A substantial part of West Africa is either in a state of crisis or potentially explosive. Among the most critical of the problems facing the subregion are the following:

1. A governance crisis that has led to the collapse or near collapse of some states and institutions. The outcome is political violence which translates into a whole generation of West African youths being schooled in the art of armed violence as a way to survive.

2.   The return of previously conquered diseases (for example tubercu-
losis), and AIDS today, threaten to decimate a substantial number of
West Africans.
3.   A continuing increase in the number of debt-distressed countries
beset by a debt crisis out of which many seem to have little hope of ever
escaping without wholesale debt cancellations.

In the last two decades West African countries have experienced diminished importance as trading partners – especially as less important export
markets at a time when economic transactions have intensified in the
relationship between the US and other regions. They have become less
important in the international financial systems, most notably as non-
creditworthy states to commercial banking systems, and to international
donors including the IMF and World Bank. They have emerged, with the
end of the Cold War, as politically marginal, due in part to their lack
of critical minerals and the end of previously exploitable great power
rivalry. They have experienced a dramatic deterioration in their state
institutions and infrastructure to the point where some (for example,
Liberia, and Sierra Leone) have fallen into the category of very unstable or

Table 2.1   Merchandise trade: US exports to, imports
from West Africa ($1,000)

	1995 Exports	1995 Imports
Benin	34,026	9,776
Burkina Faso	14,614	380
Cape Verde	7,606	259
Equatorial Guinea	5,336	30,970
Ghana	167,161	119,989
Guinea	66,617	98,951
Guinea Bissau	869	40
Ivory Coast	173,200	214,078
Liberia	41,719	9,819
Mali	23,179	5,628
Mauritania	43,113	5,541
Niger	39,537	1,557
Nigeria	602,194	4,774,668
Senegal	35,832	6,744
Sierra Leone	17,952	28,478
Gambia	6,175	2,259
Togo	18,420	29,326

Source: US Department of Commerce – Bureau of the Census,
US Global Outlook 1995–2000.

collapsing states. They have become, because of their increasing weakness, insignificant players in the solution of problems related to the global agenda: poverty, the environment, wars, diseases, and so on.

Table 2.1 reveals that West Africa's trading relationship with the United States is insignificant. Now that conceptions of national security emphasize also beneficial economic transactions as opposed to only military-strategic concerns, West Africa will continue to be of marginal importance to the United States. Among the countries of the sub-region, only Nigeria is endowed with oil, a critical resource in international trade. No West African country is among the top 50 purchasers of US goods and services. Only Nigeria falls among the top 50 suppliers (exporters) of goods to the US The insignificant trade ties with the US, makes the region of little importance to overall US foreign policy. Except for Nigeria's oil, the countries of West Africa are characterized as small markets, unstable economies, and non-critical raw material exporters. These are factors that will continue to characterize the sub-region's economic status to the United States.

Policy Review and Reorganization

Based on the stated objectives of US foreign policy, and the multiple problems that confront many West African states, what has been US foreign policy towards West Africa in this post-Cold War era? Up to 1989 US policy can be explained easily enough by the political continuity fostered by the Cold War rivalry. However, once the Cold War came to an end the US would address its costly domestic problems, and also to reaffirm the need for a meaningful foreign policy towards the entire continent.

A seven-year time frame is quite a sufficient basis for evaluating the emergence of US policy towards the sub-region of West Africa. For American policymakers, West Africa fits squarely into a hierarchical conception of policy. Certain entities like Nigeria are a resource which must be nurtured. Nigeria is all the more important politically because it constitutes the most populous nation in Africa, and is the world's seventh largest oil exporter. The US gets about 10 percent of its imported oil resources from that country. But with the turbulence and potential volatility of the sub-region in the context of post-Cold War changes, the 1990s US administrations of Bush and Clinton seem to have no solid intellectual framework around which a coherent West African policy would develop. A United States policy towards the francophone countries, for example, is absent because the US usually defers to France.[18]

The US knows that France, as an integral member of the West, can count on stable states like Côte d'Ivoire or Senegal. As Daniel Bourmand has underscored:

> France can count on the support of African states as long as it refrains from direct involvement in events within the African states. For France all leaders are acceptable, from Senghor to Bokassa, from the most liberal to the most tyrannical, on condition that they remain faithful to the metropolitan power and its interests. French power in Africa is encased in a simple code: Every head of state knows that everything will be pardoned as long as they submit to the will of France[19]. To a large extent, France's status as a middle-ranked power is dependent on her long-established access to the cheap raw and strategic materials of African states. The US therefore, values the stabilizing influence that France projects in the region. However, the US is beginning to realize that France's substantial presence and frequent meddling in Africa is not helping decisively to alleviate the region's perpetual security problems.

In this regard, the US has taken the initiative to urge African states to establish an all-African Response Force of some 20,000 soldiers to be called upon to intervene in Africa's conflicts, protect refugees, feed the starving, and prevent genocides. This initiative could be interpreted as a significant new foreign-policy development on the part of the US because if successful, it would (1) encourage Africans to learn to deal with their own numerous sub-regional conflicts; (2) constitute an endorsement of ECOMOG-like efforts, recently demonstrated in Liberia, to limit the extent of civil wars within states; and (3) help to strengthen the new democracies of the region and protect them from ethnonationalist violence.

Non-commitment and Multilateral Appropriation

Current United States policy toward West Africa is mostly one of non-commitment and multilateral appropriation. Partly a result of the changed global security system, this policy posture has come to characterize US foreign policy firmly, even to the point where warnings of Africa's increasing marginalization are getting louder than ever before.[20] Many observers see US continued disengagement from West Africa, and overall peripheralization of the sub-region as key factors in emerging US West African policy. In Liberia, due to the lack of a vigorous and committed policy of peacemaking, the US has lost many of its assets such as the giant VOA transmitter at Careysburg, as well as the intelligence listening post at Brewerville. In Sierra Leone and Senegal, its port access is being increasingly hampered by rebel efforts.[21] If the turbulence and volatility of the sub-region continues to increase more and more, US access will be denied due to widening instability. Just as rhetoric and ambivalence have come to characterize US Africa policy in general, so are the relations between the US and West Africa in the post-Cold War era based on collective appropriation and/or detachment. In the area of peacekeeping, the Somalia fiasco in which 18 American soldiers were killed in 1992, has further convinced the Clinton Administration to reject unilateral peacekeeping missions to West Africa's troubled states. A US response to any West African crisis is often part of the general international response under the umbrella of the United Nations, or as part of financial assistance to encourage demobilization, democratization, and reconstruction in war-torn states.

The combined effects of American retrenchment, ambivalence, and acknowledgment of European tutelage translate into an oscillating US West African policy, and the allocation of meager resources to the sub-region for problems that demand proportionately more resources in order to achieve the desired goals. The end of the Cold War and the growing marginalization of Africa in general almost certainly means that West Africa will receive less future assistance in dollar terms. The African Country Program Budget Proposal for FY 1996 had many West African countries receiving meager sums of foreign assistance.[22]

In general Africa is now bound to receive far less aid just as its problems are mounting. The three bills (HR 1561, HR 1562, and HR 1563) introduced by Republican Benjamin Gilman in May 1995 proposed to abolish US Agency for International Development (AID), cut funding for the Development Fund for Africa to $529 million (a reduction of 35 percent), and cut funding for the Africa Development Foundation to $10 million. Moreover, voluntary assistance for peacekeeping operations, including funding for the African Regional Peacekeeping Program, and for the OAU, would be cut to $50.4 million, a reduction of nearly 50 percent.[23] All bills concerning the reorganization of foreign affairs do not portend well for foreign aid. Africa stands to lose the most in any reduction of foreign assistance. And West Africa is more likely to suffer the most because of its civil wars, its need of peacekeeping operations, and financing requirements for reconstruction after the conflicts have ended. The burgeoning crises of West Africa – debt, collapse of commodity prices, inflation, unemployment, political upheaval, and massive health problems, among others – are occurring in the context of major structural reorganization of international relations that include massive reductions in foreign aid to Africa in general. The proposed US aid allocation for FY 1996 for the 16 countries of West Africa was modest. The recipients of the highest total aid was proposed to be Ghana ($56,938 m), Mali ($39,804 m), Senegal ($32,442 m), and Niger ($31,594 m). The entire sub-region of 16 countries were to receive about $277,000 m.[24] The insignificant amount of US aid to this region is an indication that the US is not prepared to commit itself seriously to political stabilization and the moderation of conflicts in the sub-region.

Regarding another aspect of US-Africa policy, the Bush Administration clung to the Cold War tendency of tolerating 'corrupt and dictatorial leaders', such as those of Liberia, Nigeria, Sierra Leone, and Togo. However, since 1992 election speeches and subsequently policy postures have emphasized 'democracy'. *National Security Review 30* stressed the importance of the focus on democracy: 'assisting [African] countries making progress toward democracy, better governance, and improved human rights with their defense is a legitimate aim of military aid'.[25] However, the US policy inclination of deferring to Europeans, non-commitment, and dependence on multilateralism, may not necessarily be beneficial for West Africa. In some instances US demands for democratization brings up demands for secession or encourages civil wars. As a result, current US policy, for example, is not very helpful for Sierra Leone, now experiencing a major civil conflict, or Senegal, plagued by a rebel insurgency in its Casamance region, or for Nigeria, in the grip of a military oligarchy that openly violated promised democratic reforms in 1994.

Far from ameliorating civil conflict, the post-Cold War situation and the process of democratization has exacerbated or enabled societal cleavages – in particular ethnic conflicts, an insurgency against the state (Sierra Leone), until the election a stalemated situation among warring factions (Liberia), overthrow of one of Africa's longest 'democratic' regimes (Gambia under President Jawara), ethnoregional-based regime intransigence (Togo), and the tenacity of a northern-dominated, self-perpetuating military oligarchy (Nigeria). Overall, the US has often only reacted to developments in the West African sub-region. Some would prefer a more proactive role, or a more direct involvement in conflict resolution efforts in particular. In 1996, social and political disintegration in various parts of Africa, prompted calls from the Clinton Administration for intervention through peacekeeping operations by an all-African force. In this initiative, the US is supporting the international community's growing emphasis on regional solutions for regional problems. It is also promoting regional stability to be enabled by an all-African force capable of a variety of humanitarian and peace operations in conformity with the Clinton Administration's document, *National Security Strategy of Engagement and Enlargement.*[26]

The multilateral penchant of US foreign policy toward West Africa is seen in the calls in 1990 for a linkage of economic assistance to good governance and democratic progress. These calls were simultaneously made by the European Union, certain multilateral institutions, and USAID's Democracy Initiative.[27] Participation, civil and political rights, and private enterprise are the themes that are directed at the West African states.

However, these demands ignore the realities in West Africa. The democracy being called for is Western-based political liberalization in time, tune and tempo. Moreover, the Asian 'tigers' once referred to as worthy of emulation are in many ways advanced but autocratic societies and inappropriate models. The question becomes, is the US willing to support Western-style democracies materially but within the context of very small and fragile economies in West Africa? Based on the evidence of the meager foreign aid being disbursed, and the prospects of further radical reduction in foreign assistance, it does not seem that the US is prepared to commit adequate funds to the region in order to support its stated political objectives.

The way in which American policy is confronting the weak economies of West Africa is through the limited, and now doubtful, effect of Structural Adjustment Programs (SAPs). US economic attention to the sub-region has therefore been limited to the combined pressures from the World Bank and IMF for the adoption of these adjustment programs to correct some of the distortions of prices, the lack of economic incentives, and the high cost of centralized bureaucracies, among other expectations. However, a great deal of skepticism has surrounded the utility of these conditionalities. Many in Africa believe they may even be impeding the search for an effective solution to the complexity of African economic problems.[28] Virtually all West African states, although politically fragile, have submitted to these conditionalities thereby deepening the immediate economic suffering of the masses. In other words, the US demands for free market policies and the imposition of SAPs may be serving to aggravate poverty and intensify the wave of protests against even democratically elected governments. Nigeria, Sierra Leone, Ghana, Mali, and others have experienced domestic problems related to the imposition of SAPs. *National Security Review 30* even underscored the linkage problem: 'political liberalization will continue to clash with economic reform – unrest caused by economic austerity measures has led some regimes to slow the political and economic reform process'.[29]

The United States is largely perceived as the main agent behind the externally derived pressures of economic austerity. The theory of SAPs is not viewed by Africans as yielding positive practical economic results and with a positive impact on democratization. The detrimental effects of SAPs and the supposed US role in it was emphasized in the CODESRIA May 1992 conference on 'Democracy and Human Rights in Africa: The Internal and External Constraints.' Its concluding point underscored the problem: 'The African people have demonstrated a strong will and determination to democratize their societies but – their efforts are being eroded by the constraints emanating from the SAPs imposed by Western powers and their local allies.'[30]

The conference was concerned in particular by the transformation of the world into a unipolar system in which the interests of *Pax Americana* could become a major stumbling bloc to democratization. Thus, the emphasis on SAPs and economic reform creates contradictions, and anomalies in US policy towards West African countries. The consequence has been austerity measures that have produced riots in several countries, leading to further repression by the state. Overall SAPs may generate distrust between the state and society, and deeper inequality between segments of society.

Multilateralism and Peacekeeping

The post-Cold War US administrations have embraced two concepts that have become popular since the end of the Cold War: multilateralism and peacekeeping. At one level there is nothing new about multilateralism. Throughout the Cold War period it had been US policy to summon as much international support as possible to spread the burden of peacekeeping and humanitarian action to as many partners as possible. The difference, however, is that in the post-Cold War era, the US has become more dependent on the UN in particular to deal with the major problems in Africa. To Africans the impression this conveys is that of a non-committal attitude camouflaged by a heavy dosage of rhetoric about Africa's importance to the US.

The form of multilaterialism that seems to be in operation in the case of West Africa is that of shifting the responsibility for action onto the United Nations; however, the UN without significant US or Western support is greatly limited in terms of capabilities and effectiveness. For its critics, such a reliance on the UN avoids direct US intervention to help alleviate some of the sub-region's problems. Multilateralism – cooperative action via the UN – can be effective and achieve quick results if the US or a small number of core (powerful) members have a common view of the threat to their vital interests posed by West Africa's problems.

In the case of West Africa's mounting problems and tendencies towards fragmentation, multilateralism (burden shift to the UN) is not going far enough. The problem lies in the fact that the US is not willing (especially after the Somalia fiasco) to engage its forces and put its personnel at risk except on behalf of unambiguous vital national interests, as determined by the American administration. Unless a US president can present a military mission as serving strategic US national interests, the public will not support even low levels of casualties. Thus, we see US non-involvement in the case of war-ravaged West African states, predicated on the lowest common policy denominator.

Overall, since the end of the Cold War, the US has merely reacted to developments in West African conflicts. *National Security Review 30* specifically advocates a continued US military role in conflict resolution, including a 'US military presence which will provide direct personal contacts with African military leaders ... outside US military training and exercises'. This emphasis on conflict resolution also includes a strong support for regional and OAU peacemaking and peacekeeping operations, which, it is argued, 'offer a less costly alternative to UN operations'. George Bush in 1992 approved a US contribution of $500,000 towards an OAU peacekeeping operation in Rwanda as a first step in the process.[31] However, the lesson of 'Operation Restore Hope' in Somalia suggests that in the area of conflict resolution, the US will now shy away from active military involvement to ensure peace. Thus as in Liberia and Sierra Leone, any new civil wars that may erupt in West Africa will unlikely receive the blessings of direct US military intervention. West Africa does not pose any strategic threat to US interests which would demand a strong US response. During the 1992 presidential campaign, Clinton advisor Madeleine K. Albright (subsequently US Ambassador to the UN and Secretary of State), said that candidate Clinton favored a strong US force 'to defend American interests multilaterally, if possible, unilaterally if necessary'.[32] But West Africa does not have any strategic US interests to impel the US to embark on a strong response to destabilizing situations.

Liberia's waning geostrategic importance led to America's loss of interest in the country, to the extent that America was not even motivated enough to put the Liberian situation before the UN Security Council. Since the putative 'tutelary power' did not take the initiative to bring the issue before the UN, no other Council member sponsored the resolution that would have led to a UN peacekeeping operation in Liberia. Short of direct military intervention, the US has been involved diplomatically in seeking solutions to the conflict in Liberia. But the conflict in Sierra Leone has been ignored even at that lesser level. In both countries, the US had not indicated a willingness to do anything more than quietly urge the parties to the various conflicts to go to the bargaining table, while doling out meager resources that contribute little to the alleviation of violence. Thus, in the new national security strategy of the US, humanitarian intervention short of direct US military intervention seems to have assumed the preeminence in foreign policy. The US seeks, for example, 'A stable and secure world, where political and economic freedom, human rights and democratic institutions flourish.'[33]

With the end of the Cold War, the US is now largely free from the need to pursue policies that trade off democracy and human rights against any preoccupation to contain a threatening ideology. Accordingly, US policy towards West Africa is based on (1) as limited expenditure as possible; (2) certain regional issues taking a back seat to other more important parts of the world; and (3) any policy initiatives being as uncontroversial within the US as possible.

Transnational Threats

Because US policy to West Africa is noncommittal, largely inactive, and multilaterally-based, it is mainly reactive to externally derived threats to US economic and social well-being. One of these reactions in the case of West Africa concerns what is referred to as 'transnational threats' – developments that could spillover directly or indirectly to the United States. Transnational threats from West Africa could take the form of increased transfers of drugs (heroin, marijuana, cocaine, and so on) from the Far East or South America through Nigeria and then into the United States. Another transnational threat is the cross-boundary diffusion of epidemics like AIDS and the new ebola virus. Yet another type of transnational threat has been referred to as 'disruptive migration', emanating from deteriorating socio-economic conditions, political strife, or serious natural disasters.

Nigeria may be developing into what Washington sees as 'rogue states', 'backlash states', or 'outlaw regimes' because of its blatant disregard for human rights and for world opinion. The US preoccupation with Nigeria is to apply selective pressure that would eventually transform that country into a constructive member of the international community. Anthony Lake, former Assistant Secretary of State for African Affairs, speaking about Nigeria said that 'It is very, very important to us that we do everything we can to keep Nigeria from falling apart. The possibility I've seen created in Nigeria, a kind of Angola writ large, is terrifying because we have tremendous interests in Nigeria – I mean, beyond the human dimension of it – about 10 percent, I believe, of our oil imports comes from Nigeria. About 40 percent of the heroin we seize coming into the United States comes from Nigeria . . .'[34]

American concern about the eruption of widescale conflict in Nigeria is based on several factors. First, any outbreak of serious conflict in Nigeria could create an enormous demand for humanitarian intervention by the US. The US is the only country with the capability to transport efficiently large quantities of food, medicine, and other relief aid over long distances to remote parts of the world. Second, a serious internal conflict in a very populous country like Nigeria would definitely result in the movement of large numbers of refugees into neighboring countries, overwhelming their capability to absorb them. This in turn could undermine the stability of these countries and in turn create additional demands for humanitarian assistance. Rwanda's refugee problem in Zaire is a relevant precedent. Third, the collapse of a major state like Nigeria would impede or block America's established access to oil and other economic resources Nigeria supplies. It would also block the ability of the US to restrict criminal trafficking from the region.

Currently, US policymakers are increasingly concerned about the growing role being played by Nigerian drug traffickers in the transportation of heroin to the US. Robert S. Gelbard, Assistant Secretary of State for International Narcotics and Law Enforcement Affairs, at a hearing on Nigeria held by the House Africa sub-committee said that 'Nigeria has emerged as one of the most significant drug trafficking hubs in the world, and that the rest of Africa looms as a potentially major center for narcotics production, trafficking, and money laundering in the future'.[35] Similarly, in September 1994, R. James Woolsey, Director of the CIA, declared before a conference on Global Organized Crime held in Washington, DC, that Nigerian criminal gangs were 'responsible for shipping 30 to 40 percent of all the heroin coming into the US'.[36] The Clinton Administration's response to Nigeria's role in drug trafficking was to deny certification to Nigeria for the receipt of all humanitarian assistance, in Fiscal Year 1994/5. This decision was in accordance with a provision of the Foreign Assistance Act, whereby the President is required to certify that recipients of American aid are cooperating with international efforts to control the drug trade. The US government had also voted against all Nigerian requests for loans from multilateral banks for the past two years. This decision is in accordance with a provision of the Foreign Assistance Act, whereby the President is required to certify that recipients of American aid are cooperating with international efforts to control the drug trade. The US government had also voted against all Nigerian requests for loans from multilateral banks for the past few years due to the subversion of democracy and drug-trafficking offenses. Finally, a potentially major internal armed conflict in Nigeria could fuel the flow of arms, not just into Nigeria, but into neighboring countries as well. This has the potential of spilling Nigerian instability into neighboring countries. For example, the Liberian conflict soon flowed over into Sierra Leone, Ivory Coast and Guinea. The result has been an increase in criminal violence and the complete breakdown of law and order in the hinterland of Sierra Leone, and has introduced tensions with the other countries of West Africa.

Nigeria has recently been the only country in the West African sub-region to cause the US to make some strong policy pronouncements. The US response to the most recent reversal of democracy in 1993, followed by the December 1995 hangings of Ken Saro-Wiwa and other Ogoni activists, was a mixture of caution and tough diplomatic measures against the Nigerian military regime of Sani Abacha. The response stopped short of blocking oil supplies from Nigeria. Curtailing oil from Nigeria could trigger economic problems in the US because the US already has in place oil embargos against other major suppliers such as Iran, Iraq, and Libya.

The US, since 1993, has registered its displeasure against Nigeria by implementing the following:[37]

1. Cancelled $11 million in assistance that had been earmarked as budgetary support to Nigeria's Ministry of Health.
2. Terminated all other development aid, except that channelled through NGOs.
3. Suspended all government-to-government assistance and training, except for counter-narcotics related training.
4. Denied all exports of military articles and services to Nigeria.
5. Expelled the Nigerian military attaché from the US, and withdrew the US security assistance officer from Nigeria.

Following the Ogoni executions, the US has refused backing for any IMF assistance and debt-relief for the country. The International Finance Corporation, the World Bank's private sector affiliate, has pulled out of a $3.8 billion liquefied natural gas project. In sum, the US has engaged in varied punitive measures but avoided any tougher actions against the (late) Abacha regime that could precipitate nationwide instability within Nigeria, or spawn disruptive economic effects within the US.

Rhetoric versus Vigorous Application

It is to the credit of the two post-Cold War US administrations that they have not taken the neo-isolationist view that the US has no interest in West Africa that warrants the use of any direct intervention. But they have failed to explain which specific American interests are worth intervention or to justify their inaction convincingly in relation to those West African countries that are in need of external military involvement.

The prospect that the US will pursue declared policy towards West Africa with sufficient determination or skill has been jeopardized to some extent by setting foreign policy goals that are perceived to be mere moral rhetoric but which demonstrate that this region is not central to US interests. Perceived American non-commitment towards the 'marginal' or 'non-critical' issues in West Africa, for example, could lead to a misjudgment of US intentions regarding the more 'salient' issues or interests in the sub-region.

Additionally, the pattern of espousing policy rhetoric with no attempt at vigorous implementation may cause the West African states other than Nigeria to question whether any of their members are considered strategically important in US foreign policy. By engaging in well-meaning rhetoric without comensurate commitment, the US runs the risk that it will not be capable of responding later to truly 'vital' developments, and that any commitment would be presumed half-hearted in any case. Political credibility, at least, is far too important to be dissipated, even though that has often been the outcome in US relations with marginal states, such as those in West Africa.

By way of example, instead of engaging in a great deal of rhetoric but in what many would consider only mild diplomatic responses against Nigeria's Abacha regime in 1993, the Clinton Administration from the outset could have tightened or expanded sanctions against Nigeria, including the termination of oil imports from that country, and intensified its public support for the late Chief Abiola, the presumed winner of Nigeria's 1993 Presidential elections. One would expect that the US would have decided to make of Nigeria (as it did of Haiti) an example of a way to deter antidemocratic tendencies elsewhere in the sub-region. Accordingly, it could also have extended a particularly warm embrace (as it did with Haiti's President Aristide), complete with Oval Office visits, around a democratically elected leader whose opportunity to lead Nigeria were suppressed by the military. The US could easily have arranged for alternative supplies of oil from other countries. Even if such action would have failed to persuade Nigeria's military government to hand over power to civilians, at least, it could have sent a very strong message to other potential unelected disrupters of democratization.

Although the advancement of democracy only does not constitute a salient reason for intervention, nonetheless the trend in West Africa toward democracy and market-oriented policies needs to be more energetically supported. Democratic regimes by nature tend to be effective at achieving economic growth, managing societal differences, and tolerating the effects of external dependence. Assisting democratization in the long run could stimulate the flow of external resources needed to resume economic growth in those countries, addressing the debt crisis, and stabilizing international trade flows to the benefit of the US.

A new sense of US power and purpose would enhance the American democratization mission and concentrate it on countries that seem more amenable to Western-style political liberalization. In West Africa, such countries would likely include Benin, Mali, Ghana, Senegal, and the Ivory Coast. Whether the US understands the difference between trying to create democracy in a barren setting, or reinforcing democracy where it already exists, will be a test of maturity. But it seems undeniable that American interests in the long run are more likely to be advanced by supporting democratization rather than pursuing a policy of benign neglect.

Conclusion

Overall, US West African policy tends to be one shaped by changing geostrategic concerns: the search for security. Since the end of the Cold War, deferment and dependence on multilateralism appear less the cause than the consequence of the outcome of a mostly noncommittal policy. These also account for the inverse relationship between strong rhetoric and the low level of actual support in the sub-region of West Africa. The end of the Cold War has brought about (1) a transformation in the structural relationship underlying the strategic and ideological concerns of the great powers; (2) a decline in the ability and willingness of the US to actively shape and direct events in Africa; and (3) a growing marginalization of many weak developing countries. These developments have generated a US foreign policy strategy of cautious involvement. It is a strategy predicated on following the path of least resistance: urging developing countries to eliminate policies that are both politically and economically counter-productive; taking steps that cost little or have the least likelihood of incurring American casualties; aggressively combating transnational threats; using well-tested conservative policy *modi operandi*; and avoiding uncertainty and lofty political hurdles. It emphasizes the relatively modest steps needed to weave American concerns into a fabric of mainstream foreign policy towards the region.

United States West African policy is thus a policy based rhetorically on moral purpose that in practice does very little to alleviate the immediate suffering and misery in the sub-region, and is therefore largely tantamount to a non-policy posture. This translates into America's disengagement from most of Africa which has undermined efforts to bring West African conflict situations before the United Nations Security Council, where the US could sponsor greater resolutions that would facilitate UN action. This US inaction is all part of the fear by the US foreign policy establishment of overcommitment and overextension in attempting to resolve African conflicts. The combination of fear, inaction, caution, dependence on the UN, and deferring to other powers, has generated a US West African policy based on a refusal to incur large expenses (except in Liberia) in the sub-region; a refusal to be entangled in the sub-region's conflicts because they are not central to US national security; and a tacit understanding that military commitments to the region might create adverse political controversies within the United States.

Notes

1. The many problems of the West African sub-region are sufficiently documented in recent issues of the well-known periodicals, *Africa Confidential*, and *West Africa*.
2. For details on the different roles imputed to Africa, see William J. Foltz, 'Africa in Great-Power Strategy', in William J. Foltz and Henry S. Bienen (eds) (1985) *Arms and the African: Military Influences on Africa's International Relations* (New Haven: Yale University Press); and Kent H. Butts (17 March 1993) *The Department of Defense Role in African Policy*, Strategic Studies Institute, US Army War College (Carlisle Barracks, PA).
3. Halford J. Mackinder (1904) 'The Geographical Pivot of History', *Geographical Journal*, 23 (4); and Nicholas Spykman (1944) *The Geography of the Peace* (New York: Harcourt Brace).

4. On the strategic importance of Africa's minerals, Gregg Herken (1982) *The Winning Weapon: The Atomic Bomb in the Cold War, 1945–1950* (New York: Vintage Books); and Oye Ogunbadejo (1985) *The International Politics of Africa's Strategic Minerals* (Westport, CT: Greenwood).

5. For more details, see Lisa Alfred (winter 1993) 'Africa in US Foreign Policy', *ACAS Proposed Agendas for Scholars of Africa*, nos 38–9.

6. See, for example, 'Plus Ça Change' (14 June 1991) *New Statesman and Society*, 18–20.

7. Daniel H. Simpson (1994) *US Africa Policy: Some Possible Course Adjustment* (Carlisle Barracks, PA: Strategic Studies Institute: US Army War College).

8. For further details on the politics of US Soviet rivalry in Africa, see Gerald Bender (1985) *African Crisis and US Foreign Policy* (Berkeley: University of California Press).

9. See *Report by Vice-President Richard M. Nixon to President Dwight D. Eisenhower on the Emergence of Africa* (7 April 1957), 635–40.

10. See, for example, Arthur M. Schlesinger, Jr (ed.) (1983) *The Dynamics of World Power: A Documentary History of United States Foreign Policy, 1945–1973: The United Nations and Sub-Saharan Africa*; (New York: Chelsea House), 737.

11. See, Committee on Foreign Affairs (1970) *Report of Special Study Mission to West and Central Africa*, 29 March to 27 April 1970, by Rep. Charles C. Diggs, Jr, Michigan, Chairman, US House of Representatives, Subcommittee on Africa (Washington, DC), 3–6; 9–10.

12. See, Schlesinger, Jr (1983), 731–8.

13. See, for example, Robert Legvold (1970) *Soviet Policy in West Africa* (Cambridge: Harvard University Press), 122; Courtney Sheldon (26 May 1961) *Christian Science Monitor*, 3; and William Atwood (1967) *The Reds and the Blacks* (New York: Harper and Row), 37–41.

14. For evidence of Kwame Nkrumah's growing radical anti-Western rhetoric, see *Ghana's Evening News* (4 February 1964), 1; and *Ghanian Times* (21 March 1964), 2.

15. For more details see Mark August (14 February 1996) 'Niger just last example of sad trend', *Tampa Tribune*, 4, Nation/World.

16. In *American Policy Toward Africa in the 1990s*, as quoted in *ACAS Proposed Agendas for Scholars of Africa* (winter 1993), nos 38–9, 5.

17. American Policy Toward Africa in the 1990s, as quoted in *ACAS Proposed Agendas for Scholars of Africa*, 5.

18. For details on the US role in Africa, see Christopher Coker (July 1982) 'The Western Alliance and Africa: 1949–1981', *African Affairs*, 819 (324), 319–35.

19. Daniel Bourmand (1995) 'France in Africa: African Politics and French Foreign Policy', *Issue: A Journal of Opinion*, 23(2), 61.

20. See, Tim Wall (April 1992); 'Soviet Demise Brings Africa New Challenges', *Africa Recovery 6*; and (1988) 'US Aid to Sub-Saharan Africa: Challenges, Constraints and Choices', *CSIS Significant Issue Series*, 10, 16.

21. See, for example, Lisa Alfred (winter 1993) 'Africa in US Foreign Policy', *ACAS Proposed Agendas for Scholars of Africa*, 17.

22. See, among others, Anthony Lake (29 June 1995) Assistant to the President for National Security Affairs, 'Sustained US Assistance to Africa', remarks to round table on sustained US assistance to Africa, Trans Africa Forum,

Washington, DC: in US Department of State *Dispatch* (Bureau of Public Affairs) (3 July 1995), 6(27), 537.

23. *Africa Policy Report* (31 May 1995) 'The Clinton Administration Budget Proposal for FY 1996 and Africa', published by the Africa Research Project, no. 2 (Washington, DC), 1–5.

24. As calculated from data in ibid., 2–3.

25. 'American Policy Toward Africa in the 1990s', as quoted in *ACAS Proposed Agendas for Scholars of Africa* (winter 1993), 5.

26. For further details on the background and limitations of such a peacekeeping force, see William D. Gajusz and Kevin P. O'Prey (October 1996) 'An All-African Peace Force', *Strategic Forum*, no. 86, 1–4.

27. See, for example, World Bank (1989) *Sub-Saharan Africa: From Crisis to Sustainable Growth. A Long-Term Perspective Study* (Washington, DC); and US Department of State (October 1991) *Sub-Saharan Africa and the United States.*

28. For details, see Bjorn Beckman, 'Empowerment or Repression?' The World Bank and the Politics of African Adjustment', in Peter Gibbon, Yusuf Bangura, and Arve Ofstad (eds) (1992) *Authoritarianism Democracy and Adjustment: The Politics of Economic Reform in Africa* (Upsalla: The Scandinavian Institute of African Studies).

29. 'American Policy Toward Africa in the 1990s', as quoted in *ACAS Proposed Agendas for Scholars of Africa* (winter 1993), 7.

30. *Democratization and Human Rights in Africa: The Internal and External Constraints* (CODESRIA: Dakar; SAPES TRUST-Harare, May 1992).

31. 'American Policy Toward Africa in the 1990s' as quoted in *ACAS Proposed Agendas for Scholars of Africa* (winter 1993), 7.

32. Jim Cason and Bill Martin (winter 1993) 'Clinton and Africa: Searching for a Post-Cold War Order', *ACAS Proposed Agendas for Scholars of Africa*, nos. 38–9, 4.

33. Anthony Lake (15 February 1995) National Security Adviser. Briefing at the Center for Strategic and International Studies and as quoted in *Africa Policy Report*, (30 September 1995) no. 3, published by the Africa Research Project (Washington DC).

34. *Africa Policy Report* (30 September 1995), no. 3, 8.

35. Ibid.

36. Ibid.

37. George E. Moose (20 Jul 1995) Assistant Secretary for African Affairs, Statement before the sub-committee of African Affairs of the Senate Foreign Relations Committee, Washington DC; in US Dept. of State *Dispatch* (31 July 1995) Bureau of Public Affairs, 6(31), 604.

3
Central Africa: Initiatives Deferred
Raymond W. Copson

Introduction

Central Africa, defined here to include Cameroon, Chad, Central African Republic (CAR), Congo (Brazzaville), Congo (Kinshasa, formerly Zaire), Equatorial Guinea, and Gabon, as well as Rwanda and Burundi, is a vast region that cannot be ignored by the United States or other major actors in international politics or the world economy. It is a region of immense wealth in terms of natural resources, including environmental riches as well as strategic metals and petroleum. The former Zaire, renamed the Democratic Republic of Congo (DROC) by Laurent Kabila, the victorious rebel leader, in May 1997, lies at the heart of Central Africa and is the key to its future. The fortunes of this populous, resource-endowed country affect the economic and political prospects not only of the entire Central African region, but also of much of southern, eastern, and western Africa. Congo has been a focus for US policy since 1960, both as a stake in the Cold War and because events there can affect so much of the continent.

US attention has also been drawn repeatedly to Central Africa by humanitarian disasters arising from poverty and ethnic divisions. The anti-Tutsi Rwanda genocide that broke out in April 1994 was the greatest such disaster, with 500,000[1] to 1 million lives lost, and it set the stage for the suspected deaths during 1996 and 1997 of large numbers of Hutu refugees in Zaire. The dimensions of the killing of Rwanda refugees during the Zaire civil war may never be known but it seems possible that the number will be in the tens of thousands.

Several of Central Africa's countries rank among the world's poorest in per capita incomes, and this deepens international humanitarian concern over the region and its people. Economic liberalization, which has sparked accelerating growth in some other parts of Africa, has been halting in Central Africa at best; and in all but one or two cases, democratization has been arrested or has made only slight progress. Human rights are frequently violated by the governments of the region.

Central Africa's many problems mean that policymakers, whether in the United States or other industrialized countries, have been able to do relatively little in recent years to affect the course of events there or to advance their policy objectives. With major expenditures of economic aid and political capital, of course, the Western developed countries might have had more of an impact in Central Africa. But policy outcomes in this troubled region are problematic at best, and resources to be expended on foreign affairs in the post-Cold War era are everywhere limited. Consequently, at least until the outbreak of the Zaire civil war in October 1996, none of the principal actors in the international system were willing to launch major policy initiatives in Central Africa.

Instead, US policymakers have worked at the margins in Central Africa, encouraging incremental change with only modest expenditures of policy resources while awaiting a time when more might be accomplished. The United States has supported World Bank and International Monetary Fund (IMF) efforts to promote economic reform in Central Africa without expecting major economic gains. Policymakers have exerted limited pressures for human rights and democracy reforms – sending or withholding elections observers, for example, depending on the likely credibility of any particular vote – with limited hopes for lasting improvement. The United States has provided relief and humanitarian assistance during disasters, but policymakers have sought to avoid direct involvement of US troops in peacemaking or peace enforcement actions. The risks and costs of such deployments would be difficult to justify to a reluctant Congress and public, policymakers believe, particularly since long-term solutions to the region's conflicts and instability will not soon be found.

Nor have policymakers expected initiatives to promote trade and investment – a major Africa policy objective of the Clinton Administration – to be highly successful in Central Africa. An important address on this policy delivered by George Moose, the US Assistant Secretary of State for African Affairs, in April 1995, focused heavily on southern Africa and made reference to West Africa and East Africa as well, but did not mention the Central African region.[2]

Secretary of State Warren Christopher did not stop in Central Africa during his swing around the continent in October 1996, although, Rwanda and Burundi, site of Africa's gravest current humanitarian crisis, was an important subject in the discussions he held. US policymakers entertained some hopes that the elections promised by Zaire's President Mobutu Sese Seko for 1997 would have heralded the beginning of a turnaround in Central Africa but this did not materialize.

The United States had participated in humanitarian efforts to relieve the suffering of the Hutu refugees who fled Rwanda in July 1994. Troops from the United States and other countries were sent to the region to assist in international relief efforts for the million or more Hutu refugees who fled just across the Rwanda border into Zaire. Hutu militants, including members of the former Rwanda government and the former army, soon won control of the vast refugee camps, where they plotted a return to power in Rwanda – a situation the Rwanda government could hardly tolerate. The failure of the international community to deal with this unstable situation between July 1994 and October 1996 helped set the stage for the Zaire civil war.

Kabila's decisive victory in that war, ending the Mobutu era, may have fundamentally altered the regional political and economic situation, but this remains to be seen. If it has done so, the level of US engagement in the region will certainly increase. With a new regime in power in Congo, US economic assistance and military training could resume. Trade could expand sharply in view of Congo's potential as a supplier of raw materials and potential market for US goods, while US businesses could begin to invest heavily in expectation of an economic expansion in Congo itself and throughout the region. But these developments are contingent on the course President Kabila chooses to follow. US–Congolese ties can only strengthen if, in the immediate future, there is progress in the investigation of alleged atrocities during the civil war; and if, over the longer term, Kabila maintains a commitment to economic liberalization, respects human rights, and makes headway in democratization.

US Interests and Involvement

During the Cold War, Central Africa, particularly Zaire, ranked higher on the US policy agenda than afterward. There was little interest in the region before the Cold War, although a number of American missionaries devoted their careers to evangelizing the Belgian Congo, now Zaire – and to building schools and hospitals there.[3] Trade with the region has been negligible apart from limited oil imports from Gabon and Congo. The Central African region assumed considerable geostrategic importance in 1960, however, when Zaire unexpectedly became a major stake in the Cold War. Belgium granted Zaire independence on 30 June 1960, but when it became evident that Belgian officers would continue to command the armed forces, the army mutinied, Europeans began to flee – prompting a limited Belgian armed intervention – and mineral-rich Katanga province seceded. Prime Minister Patrice Lumumba, meanwhile, called on the Soviet Union for support in his escalating rivalry with President Joseph Kasavubu. At the urging of UN Secretary General Dag Hammarskjöld, a large UN peacekeeping force was deployed to try to stabilize the situation and cool superpower concerns, but when Lumumba was assassinated in early 1961, Soviet outrage nearly made the United Nations itself a Cold War victim. The Katanga secession came to an end only in January 1963, following a violent showdown with UN forces. Secretary Hammarskjöld had been killed in September 1962 in a suspicious plane crash over Northern Rhodesia (now Zambia) while flying to peace talks with secessionist leader Moïse Tshombe.

United Nations peacekeepers finally left Zaire in June 1964, despite continuing rural insurgencies.[4] In November 1964, US transport planes carried 500 Belgian paratroopers to Stanleyville (now Kisangani) to rescue foreign hostages held in one such insurgency. This intervention helped government forces secure the area.[5] Three US C-130 transports were sent to Zaire again in 1967 to provide logistical support to the government during an uprising by Katanga gendarmes backed by 200 white mercenaries. The State Department explained that the United States was helping the Kinshasa regime 'as it has consistently done in the past, in that government's search for stability and peace'.[6]

By this time, President Mobutu Sese Seko, a former army colonel and journalist who had seized power for the second time in November 1965, was consolidating his power, and he had just made Zaire a one-party state. In the Cold War context, his regime won the ready support of the United States, Belgium, and France – the principal Western actors in Zaire – despite the complaints of his political opponents.

Relations with Zaire grew strained under the Carter Administration (1977–81) because of the emphasis that Administration placed on human rights. Mobutu was angered by Carter's refusal to provide more than 'non-lethal' military assistance during the 'Shaba I' crisis of 1977, which saw former Katanga gendarmes, who had been exiled in Angola, attempt to seize power in Shaba Province, as Katanga had been renamed. Mobutu and others attempted to portray this crisis as part of a 'grand design'[7] for Africa on the part of the Soviet Union and Cuba, both heavily involved in supporting the Marxist government in neighboring Angola in its civil war. Carter's reluctance to intervene was an outgrowth of the Vietnam experience and an early sign of American wariness with respect to military entanglement in African conflicts. However, the 'Shaba II' uprising of May 1978 was much more serious, and in this instance Carter lent US Air Force transports, with pilots and support personnel, in support of a French-Belgian intervention that ended the rebellion. Some Members of Congress were not convinced by the Cold War justifications advanced in support of this operation, particularly intelligence reports showing a Cuban role in training the insurgents.

US relations with Zaire improved markedly in the Reagan years, even as congressional criticism of Zaire's human rights violations mounted and the Zairian economy continued to deteriorate despite sporadic efforts at reform. In 1983, Reagan described Mobutu as 'a faithful friend to the United States for some 20 years'.[8] Zaire's help with a crisis in Chad was a factor in improved relations, as was its probable cooperation on Angola later in the decade when the United States launched a publicly acknowledged covert operation in support of the UNITA resistance to Angola's government. Subsequently, Mobutu was the first African head of state to meet with President Bush in the White House.

However, the Bush years also saw the beginning of the sharp devaluation of Zaire as a factor in US Africa policy. Rising pressures for democratization within Zaire and the country's economic decay suggested to policymakers that Mobutu's regime was becoming unstable. With declining Cold War tensions, moreover, and progress toward peace in Angola, the United States had less need to cooperate with the Zairian leader than in the past. Secretary of State James Baker went to Zaire in March 1990 and told Mobutu that in the absence of reforms in human rights, the economy, and politics he risked being swept aside. Unrest in September 1991 prompted Assistant Secretary of State for Africa, Herman Cohen, to demand that elections be held in Zaire within one year.

Pressure on Zaire for reforms only increased during the Clinton Administration, as the concluding section of this chapter, on current US policy, will note. At the same time, however, the resources policymakers were willing to devote to Zaire prior to the civil war were quite limited. Zaire remained important primarily in relation to the humanitarian crises in Rwanda and Burundi, and because of fears that a comparable or even more grave crisis could develop in Zaire if the political system finally collapsed.

During the Zaire civil war, international interest in that country and in Central Africa generally began to intensify, as it seemed that a regional situation frozen for years by Mobutu's hold on power in the region's largest country might be loosening. In November and December 1996,

France pushed strenuously for a refugee protection force to be sent to north-eastern Zaire, and such a force was authorized by the Security Council. But neither France nor then UN Secretary General Boutros-Ghali could overcome US caution with respect to possible entanglement in African conflicts. President Clinton had tentatively agreed that the United States would participate, and US troops were slated to secure the airport at Goma – a task that could have led to clashes with the Hutu militants. Secretary of State William Perry said that the US force would have 'robust' rules of engagement and would be fully equipped to protect itself. Nonetheless, he insisted that 'We are not planning a mission to go in and disarm factions or to separate the military from refugees.'[9]

The hold of the militants over the camps was broken on the night of 14 November 1996, when anti-Mobutu rebels attacked, and more than 700,000 refugees eventually returned to Rwanda while thousands more fled deeper into Zaire. The Clinton Administration could no longer see a clear military objective for the proposed international force, and withdrew its support. Kabila's rebel group, the Alliance of Democratic Forces for the Liberation of Congo–Zaire (ADFL), had opposed deployment on grounds the force would provide a cover for a restoration of Mobutu's control over the region. In any event, without US backing in transport and logistics, the force could not become operational, while the few Canadian troops sent to Uganda to spearhead the deployment were withdrawn.

Chad was another Central African country that once attracted significant attention from US policymakers, particularly after 1979, when Libyan-backed guerrillas from northern Chad seized Ndjamena, the capital. Washington had long been concerned by what it perceived as the threat posed by Libyan leader Mu'ammar Qadhafi to political stability south of the Sahara. Although this guerrilla faction, led by Goukouni Oueddei of the northern Toubou tribe, asked Libyan forces to leave Chad in 1981, US officials were pleased when a rival northern figure, Hissein Habré won back the capital in 1982. In 1983, with Libya again active in the north, France sent troops into Chad to stabilize the Habré regime and the United States launched a military assistance program focused on the delivery and maintenance of transport aircraft and the provision of overland vehicles.[10] Press reports indicated that there was a substantial covert US role on Habré's behalf, and that this played an important part in his decisive victory over Libyan troops in northern Chad in 1987.

Habré failed to build a national political consensus after 1987, and his last years in power were marked by clan conflict and charges of human rights violations. He was forced to flee in December 1990, when another northern guerrilla leader and former ally, Idris Déby, swept into Ndjamena from Sudan. While Déby had Libyan backing, the broader Qadhafi threat seemed to be fairly well contained by this time, and the invasion failed to set off alarm bells in Washington – which was in any case preoccupied by the crisis in the Gulf region. Indeed, the receding Libyan threat and the apparent impossibility of achieving lasting policy gains in fractious Chad, meant that this country too began to lose importance in America's foreign policy as the 1990s began.

Rwanda is the third Central African country that has been a major US policy focus, in this case because of a quite recent humanitarian crisis rather than the long-term strategic considerations that drove policy toward Zaire and Chad. The latest crisis began in April 1994, when a plane carrying Rwanda President Juvenal Habyarimana, a Hutu, back to Rwanda from a regional peace conference, was shot down while approaching the airport at Kigali, the capital. The Burundi president was also on board. Outside observers suspected that Hutu extremists in the military had downed the plane in order to sabotage a peace agreement with the Rwanda Patriotic Front (RPF), a predominantly Tutsi resistance movement. Hutu leaders, however, largely blamed the RPF, and Hutu militants launched a campaign of genocide against the Tutsi, killing many Hutu moderates as well. Many observers believe that this genocide had been planned well in advance. The RPF resumed military operations in the face of the genocide and as a result of a remarkably successful offensive, it was completing its conquest of Rwanda by mid-July 1994, when more than a million Hutu refugees and the remnants of the largely Hutu Rwanda army fled into Zaire.

The Clinton Administration initially seemed reluctant to become deeply involved in the Rwanda crisis, although President Clinton spoke out against the violence and sent two special envoys to try to arrange negotiations. With the rest of the UN Security Council, the United States voted on 21 April 1994 to reduce a small United Nations Assistance Mission in Rwanda (UNAMIR), deployed earlier to oversee implementation of the earlier peace accord, on grounds that its personnel were at risk. This move was sharply criticized by the Organization of African Unity's Salim Ahmed Salim, among others.[11] The Security Council, with the support of the United States, voted in favor of a 17 May resolution to expand the force, but critics pointed out that by this time most of the victims of the genocide had probably already perished. In any event, member states were slow to commit troops and the new force was not fully deployed for months. Members of Congress criticized the

Administration for much of the delay, including problems that arose in sending 50 promised armored personnel carriers to Rwanda. Representative Donald Payne was particularly critical of the US Ambassador to the United Nations, Madeleine Albright, whom he accused of 'participating in one stalling maneuver after another',[12] while House Africa Sub-committee Chairman Harry Johnston criticized the Department of Defense for 'dragging its feet' on Rwanda.[13] The Administration endorsed the June 1994 French deployment of a small humanitarian protection force in Rwanda, but did not volunteer US troops.

With the July 1994 refugee crisis, which overwhelmed relief organizations operating in eastern Zaire, the United States escalated its involvement by deploying 'Operation Support Hope'. Nearly 2,600 US troops participated in this operation, assisting in logistics to speed relief supplies to the region, setting up water purification facilities for the refugee camps in Zaire, and providing other relief aid. However 230 US troops actually went into Rwanda, where they kept the airport operational for round-the-clock relief flights. The operation was kept to a strict timetable and concluded on 30 September 1994.

Administration supporters maintained that overall, US policymakers responded appropriately to the Rwanda humanitarian emergency, both before and after the July refugee crisis, while minimizing risks to US personnel and avoiding long-term military commitments in Central Africa. US caution on expanding the UN force was fully justified, from this perspective, in view of the difficulties experienced by the United Nations in Somalia and heavy UN commitments elsewhere. Thus, in the face of criticism, President Clinton said on 24 July 1994, that 'We've been working [on the Rwanda crisis] since May and I have done all I knew to do.' Relief workers and others, however, felt that the United States should have made heavier commitments in Rwanda and left its military forces in the region for a much longer period.

US Assistance Trends

The principal patterns of US interests in Central Africa are reflected in the foreign assistance trends for the region during the 1985–95 decade, as shown in Table 3.1.[14] These data underline the importance of Zaire during the Cold War years and its subsequent fall from favor. The Administration's disappointment with Mobutu's failures in political and economic reform was very deep and, by 1995, Zaire was receiving no aid of any kind. Moreover, it had accumulated large arrears in the repayment of concessional loans made under the aid program in earlier years, even

Table 3.1 US bilateral assistance in Central Africa (fiscal years, millions of constant dollars)

	1985	1990	1995
Burundi	9.0	24.1	5.1
Cameroon	45.5	30.7	3.1
CAR	6.3	6.0	4.1
Chad	52.3	20.3	4.9
Congo	1.8	3.5	7.8
Eq. Guinea	2.5	1.8	.2
Gabon	2.5	2.8	2.8
Rwanda	32.0	15.6	171.5
Zaire	89.4	62.7	0
Total	241.3	167.5	199.5

Source: US Agency for International Development publications. Includes economic aid (development assistance, Economic Support Fund, food aid, and Peace Corps) as well as military aid.

though these loans had low interest rates and extended repayment periods. This made Zaire legally ineligible for most types of aid.[15]

The importance of Chad during its confrontation with Libya is indicated by the high level of assistance it was receiving in 1985. Trends in aid levels point to some minor themes in US policy, including hopes held for economic development in Rwanda before the 1994 collapse, as well as support for economic reform programs there and in Burundi. Cameroon was once a larger aid recipient because of the importance placed by policymakers on political stability in a country of some oil wealth, as well as a sense that Cameroon's resources might make respectable growth possible. With the US emphasis on democratization in recent years, high aid levels for Cameroon could not be sustained. By 1995, the assistance program in the Central African region had become largely a humanitarian relief effort, and most assistance went to Rwanda.

Regional Background

The resource endowment of Central Africa is truly impressive (Table 3.2), assuring that the region will remain of interest to the US Government and private sector despite its many social and economic problems.

The mineral resources of Central Africa include the former Zaire's cobalt, constituting half of world reserves,[16] as well as its copper, industrial diamonds, and gemstones. Gem quality diamonds are also mined

Table 3.2   Central Africa: scale and resources

	Size	Population	Resources
Burundi	Maryland	6.3 million	nickel, uranium, vanadium
Cameroon	California	13.5 million	petroleum, cement, natural gas, hydroelectric
CAR	Texas	3.2 million	some diamonds, gold, uranium
Chad	3 times California	5.6 million	natron, salt, limited petroleum
Congo	Montana	2.5 million	petroleum, cement, hydroelectric
Eq. Guinea	Maryland	0.4 million	petroleum, natural gas and condensate, hydroelectric
Gabon	Colorado	1.1 million	natural gas, manganese, petroleum, cement, uranium
Rwanda	Maryland	8.6 million	tin, natural gas
Zaire	1/4 US	44 million	cobalt, copper, diamonds, gold, zinc, hydroelectric

Principal sources: US Central Intelligence Agency, *World Factbook, 1995*; *Europa Yearbook, Sub-Saharan Africa, 1995* (London: Europa); US Department of the Interior, Bureau of Mines, *Mineral Industries of Africa, 1993*.

in the Central African Republic, as is gold, and a high-grade uranium deposit has been found there. Hydroelectric potential is quite significant in most of the region, apart from Chad and CAR. A large natural gas deposit under Lake Kivu, between Rwanda and Zaire, could promote growth throughout the area but cannot be developed while the region remains politically unstable.

The coastal Central African countries share in the massive oil wealth of the Gulf of Guinea, which will be an increasingly important supplier of the world's energy needs in coming years. Petroleum production has been a steady earner for several countries for some time, and new discoveries are giving a boost to the region's prospects. Equatorial Guinea faces a potential economic boom as Mobil Oil Corporation prepares to begin producing 40,000 barrels of oil per day at the Zafiro field off Bioko Island.[17] The political consequences of this boom on the country's repressive political system remain to be seen. The Brazzaville government, meanwhile, is expecting its oil revenues to double to about $341 million per year when Elf-Aquitaine, the French oil giant, opens the Nkossa offshore oil field.[18] Even Chad, far from the coast, could begin to experience oil-financed growth if a proposed pipeline through Cameroon to export large amounts of oil from southern Chad wells is ever built.[19] Preparations for this project by an international oil consortium, including the Exxon Corporation, appear to be on track, although construction has not yet begun.

Despite the region's natural wealth, the infrastructure of Central Africa is generally poor, and in some instances, particularly in the former Zaire, its condition has deteriorated from colonial times. Rail lines, built to extract resources, are not interconnected; and neither Chad nor CAR, both quite large, have railroads at all – although CAR has hopes of bringing the Transcameroon line to Bangui. President Omar Bongo of Gabon has insisted on pushing the 385-mile Transgabon railway deep into the interior, but it is not yet clear whether this will turn out to be an unprofitable prestige project, as donors have feared, or pay its own way with timber, mineral, and passenger traffic.

The former Zaire, the region's largest country, has only about 1,500 miles of paved road, and the general breakdown of its road system is legendary among journalists and other travellers. The telephone and postal systems of Zaire are also near collapse, greatly complicating communications. The country's elites compensate in part by relying on cellular phones. These weaknesses in transport and communication pose grave obstacles to the development of Congo (Kinshasa), but at the same time, they offer opportunities to US and other foreign firms.

Two Central African countries, Chad and CAR, are landlocked, enhancing their transport problems. CAR does conduct international trade via a thousand-mile river-and-rail route to the Congo port at Pointe Noire, but this tortuous connection is no asset in terms of economic development. Congo (Kinshasa) is almost landlocked, having only a modest port on the Zaire River at Matadi. Its minerals from Shaba province must be exported by rail and truck through ports in South Africa and Tanzania.

Central Africa's vast ecological problems are of great concern to many Americans, and significant to the future of the global environment. A vast rainforest stretches from Equatorial Guinea and Gabon across Zaire. The timber, game, and rare plant species of this forest are under threat from logging, wildcat mining, and other human activities including armed conflict. Gabon's extensive Minkébé forest, home to thousands of forest elephants, is reportedly endangered by loggers seeking its tropical hardwoods.[20] Meanwhile, rebels and refugees from southern Sudan have reportedly killed off rare rhinoceros and giraffe species in Congo's (Kinshasa) northern Garamba National Park, a UN World Heritage site.[21] Currently the fate of gorilla populations in north-eastern Congo remains uncertain after several years of turmoil. And Rwanda refugees in the area are known to have cut down large numbers of trees.

Social Climate

The social climate in much of Central Africa generally is unsettled, with much ethnic rivalry and conflict, as well as high rates of crime, banditry, and corruption. These problems inhibit the region's development and could limit US involvement in Central Africa, despite its resource endowment, for years into the future. Central Africa's social problems are in part rooted in the region's poverty, as reflected in low levels of performance on most social indicators. Three of these indicators are reported in Table 3.3, and others would paint a similar portrait.

Gabon's extraordinarily high income is an anomaly for the region, and indeed for sub-Saharan Africa generally, because of its oil wealth; but the majority of Gabon's people are poor nonetheless, as suggested by its life expectancy and literacy rate. In the situations of extreme poverty encountered throughout Central Africa, high levels of theft and robbery are hardly surprising. Nor is it surprising that customs officers, other civil servants, police, and soldiers should resort to petty bribery and extortion.

The looting that broke out among soldiers in Kinshasa, Zaire in September 1991, provoking an armed evacuation of foreigners by French and Belgian troops, reflected genuine grievances on the part of the ill-paid troops. It also reflected the general breakdown of Zaire's institutions of government, which at the time were permeated by corruption extending to the highest levels. Under conditions such as these, foreign

Table 3.3   Social indicators in Central Africa (*1989 data)

	GNP/Capita 1993 ($)	Life Expectancy 1992	Literacy 1990 (%)
Burundi	180	48	50
Cameroon	770	56	54
CAR	390	47	38
Chad	200	47	30
Congo	920	51	57
Eq. Guinea	360	48	50
Gabon	4,050	54	61
Rwanda	200	46	50
Zaire	240*	52	72

Source: World Bank (1995) *African Development Indicators, 1994–1995* (Washington, DC: World Bank), 'Life expectancy' refers to expectancy at birth. 'Literacy' is the proportion of the population over 15 who can read.

assistance agencies, foreign-based non-governmental organizations, and foreign businesses incur high additional costs that limit their involvement in Central Africa. Since 1993, Bangui, the capital of the Central African Republic, has also been subject to repeated military mutinies – a serious deterrent to visits by potential investors, let alone tourists. Poverty affects other aspects of life in Central Africa as well. The region's worst conditions are probably found in Kinshasa, where rates of infection with HIV are extraordinarily high and there is widespread malnutrition, malaria, and hepatitis. A cholera epidemic broke out in the city in 1996. Living standards have been savaged by hyperinflation, and Mobutu became so unpopular in his capital that he rarely ventured there. Around the former Zaire, the formal economy hardly functions, and the people survive primarily on the informal economy, based on wildcat diamond mining, illegal exports, petty trading, small-time corruption and bribery, and subsistence farming. Conditions are less extreme in most other countries, apart from Rwanda and Burundi, but life is not easy anywhere in the region. Periodic shortages in the markets, limited access to health care, inadequate facilities for education, and a host of other problems confront the people of Central Africa in their day-to-day lives. For US and other foreign businesses that might consider investing in Central Africa, these conditions mean that a qualified workforce, capable of meeting production timetables and Western quality standards could be very difficult to assemble. Population growth rates of well over 2 percent in most of the region, and 3.18 percent in Zaire,[22] only contribute to instability and make it more difficult for political systems to respond to social needs.

Central Africa's social problems are also borne of the complex ethnic, geographic, and in some instances religious divisions that permeate the region. The armed insurgencies that have destabilized Chad since 1965 reflect an overarching north/south divide, which is also a Muslim/non-Muslim divide, as well as myriad clan and sub-clan rivalries. Political tensions in Congo (Brazzaville) have also reflected both regional and ethnic divisions – although there are also deep personal antipathies between two of the principal contenders for power, President Pascal Lissouba and former President Denis Sassou-Nguesso. Armed clashes among factional militias took place in Brazzaville in 1993, with sporadic outbreaks in 1994 and 1995, and major new fighting as scheduled elections approached in 1997.

As in the Congo instance, politics elsewhere in Central Africa often seem only a veneer covering deeply rooted social cleavages. This was long clear in Mobutu's Zaire, despite the president's claim that he opposed tribalism. Mobutu showed a marked preference in his top political appointments for members of his own Ngbandi clan. His most reliable instrument of power, the Israeli-trained Division Spéciale Présidentielle (DSP), was reportedly manned by Ngbandi and foreigners; while the army generally had been largely purged of officers from eastern Zaire, where the civil war began. They were replaced by personnel from Mobutu's own Equateur province.

Kabila, the new president, has faced criticism in the capital for his reliance on Tutsi as advisors and soldiers. The Tutsi-dominated Rwanda government is widely suspected of providing essential help to Kabila in his conquest of Zaire, and this has opened Kabila to charges that he has imposed a foreign-dominated government on Congo. Zairian-born Tutsi are a tiny portion of the Zairian population, but are today well-represented in the Kabila regime. Foreign Minister Bizima Karaha and the Secretary General of the ADFL, Deogratias Bugera, are from this group.

The perception that Kabila's government is under the control of foreigners has been exploited by Tshisekedi wa Mulumba, a leading opposition figure and head of the Union for Social Democracy and Progress (UDPS). Tshisekedi has suffered greatly for his advocacy of democracy and human rights in Zaire over many years, but his base of support has an ethnic dimension as well, since he is a Luba from Kasai and most popular among fellow Luba. Kabila himself is a 'Balubakat' – descended from Luba of the Kasai region who went to Katanga province in the colonial era in search of jobs. The Balubakat have suffered much persecution in Shaba in recent years, and hundreds of thousands were driven out in 1992–4. Kabila has appointed members of several non-Tutsi ethnic groups to his cabinet, but it will require exceptional political skill on his part to dampen perceptions that his is a Tutsi-dominated regime, and to assure other ethnic groups that they are respected participants in the new Congolese political system. Should he fail on either account, destabilizing ethnic conflict that will deter aid and investment could well ensue.

In Rwanda and Burundi, anthropologists are uncertain as to whether there is a genuine ethnic difference between the Hutu and the Tutsi, who speak the same language and whose cultural practices are similar. Some regard the Tutsi minority as an upper caste that emerged locally, rather than accepting the more popular view that they are descendants of Nilo-Hamitic herders who entered the region several centuries ago. In any event, modern political life has been almost completely structured by the Hutu–Tutsi division, which has also affected the churches[23] and other aspects of society.

Even leaders who have sought to rise above the ethnic division find themselves compelled by circumstances to yield to its pervasive influence. Burundi's Pierre Buyoya, for example, the Tutsi military ruler who orchestrated the country's first democratic election in 1993 – and gracefully stepped down when he was defeated – found himself accepting the presidency of a new, Tutsi-dominated military regime after a July 1996 *coup*. His decision came after three years of renewed ethnic violence, and it in part reflected Tutsi fears that they were being encircled in Bujumbura, the capital, by a burgeoning Hutu guerrilla movement.

Regional Relations

Until the Zaire civil war, interstate conflict had generally not been a significant problem for the countries of Central Africa, where armies have traditionally been weak and politicians have focused on domestic issues. Cameroon faces a conflict with Nigeria over the oil-rich Bakasi peninsula, but neither government seems willing to risk a major clash over the issue. Regional tensions in Central Africa arise primarily from internal conflicts that can give rise to destabilizing refugee flows and cross-border movements by rebel groups seeking shelter. Internal conflicts also tempt leaders to blame their problems on others, as Brazzaville governments often attributed their difficulties to interference by Zaire, and it has always been difficult for the outside observer to determine, on the basis of available evidence, the degree to which such charges may be true. But certainly internal conflicts can create temptations for meddling, whether by Central African states or others – as Libya and Sudan have repeatedly meddled in Chad's domestic struggles over the years. Zaire's President Mobutu was long involved in the internal affairs of Angola, a neighbor in the southern Africa region, where he was a key foreign backer of armed resistance to the Angolan regime. Mobutu sent troops into Angola in 1975 in an abortive effort to prevent the then-Marxist Popular Movement for the Liberation of Angola (MPLA) from coming to power, and Zaire also became a conduit for arms being channeled to Jonas Savimbi's UNITA movement. The US government ended covert aid to Savimbi in 1976 under congressional pressure, but resumed a publicly acknowledged, covert aid program to UNITA in 1986. Reports indicate that Mobutu made facilities in Zaire available for this program.

Despite a 1988 regional peace accord, brokered in part by US diplomats, Mobutu's cooperation with UNITA evidently continued right up until the Zairian leader's ouster. Meanwhile, Uganda's relations with Mobutu were strained over suspicions that he was allowing rebels from

Uganda to shelter in north-eastern Zaire. Mobutu had been friendly with President Habyarimana of Rwanda, whom he allowed to be buried at Gbadolite, Mobutu's remote hideaway in northern Zaire after the Rwanda Patriotic Front took power.[24] But the RPF's relations with Mobutu were strained by Zaire's tolerance of the activities of Hutu militants in the refugee camps around Goma.

Zaire's prolonged economic decline under Mobutu, and the suspicions that were often directed against it, meant that there was little room for regional economic cooperation in Central Africa over the years. While Mobutu had his friends among other authoritarian rulers in the area, particularly Gabon's President Bongo, others, such as Uganda's President Yoweri Museveni, regarded Mobutu's Zaire as an obstacle to regional development. This factor may have played a role in persuading Museveni to assist Kabila's rebels – as he is widely alleged to have done. Rwanda is also suspected of lending substantial support to Kabila, as is Angola, which evidently facilitated the return of the Katanga gendarmes, and their sons and grandsons, to join in the fray on Kabila's behalf. UNITA troops, meanwhile, are believed to have stiffened the Zaire army's brief resistance to the ADFL advance in a battle around Kenge on the approach to Kinshasa.

Rwanda, Uganda, and Angola denied substantial involvement in the Zaire conflict for some time, but diplomats did not give credence to these denials. On 9 July 1997, however, the *Washington Post* published a report and interview in which Paul Kagame, the Rwanda Defense Minister and Vice-President, revealed an extensive Rwandan role.[25] According to Kagame, the most powerful figure in Rwanda, his government planned and directed the Zairian war and sent forces to participate in the capture of Kisangani, Kinshasa, and other cities. Rwandan officers were present at all levels of command during the conflict, Kagame said.

A government is now in power in Congo (Kinshasa) that is quite friendly to Uganda, Rwanda, and Angola – as well as Burundi, and regional economic cooperation could soon begin to increase. US interests in the region could well benefit as cross-border transportation and communications improve, and regional trade grows. Whether these gains will be realized, however, depends on Kabila's success in maintaining political stability in his country.

Political Trends

Everywhere in Central Africa, pressures for democratic reforms and multiparty political systems began to mount in the later 1980s, provoking a number of major political changes in 1990 through 1993. These pressures were in part domestic, reflecting a reaction to long years of economic deterioration and one-party, authoritarian rule. Pressures for reforms also arose from the Western donor community, including the United States under both the Bush and Clinton Administrations. These pressures reflected the increasing impact of human rights concerns on Western policies as well as a growing conviction that repressive regimes were potentially unstable because of the resistance they aroused. Donors came to see repressive systems as economically dysfunctional because they were unresponsive to the needs of diverse groups in society.

Political change in Central Africa was facilitated by the end of the Cold War, which helped to convince the donors that authoritarian governments need no longer be indulged on national security grounds. Moreover, the collapse of Marxist regimes in Eastern Europe and the Soviet Union tended to delegitimize authoritarian government generally, and it removed external sources of potential sympathy and support for Africa's authoritarian rulers. President Mobutu, in making his first concessions to democratic forces in April 1990, was probably motivated in part by the revolution in Romania, which had been a close foreign backer. According to one source, 'The fall of Ceauşescu, vividly presented on Kinshasa television, reportedly made a strong impression upon Mobutu, whose announcement of democratization followed shortly thereafter. Popular humor in the capital speculated upon the future of "Mobutu Sesesescu".'[26]

In Congo (Brazzaville), a committee to examine the implications of the changes in Eastern Europe for the one-party state was appointed in February 1990, preparing the way for an extraordinary congress of the ruling Marxist party in December.[27] At this congress, the party voted to disassociate itself from Marxism-Leninism and approved the introduction of a multiparty system in January 1991.[28]

The domestic sources of political change, however, were extremely powerful and outweighed the impact of events in the European communist world. In Brazzaville, it was a general strike called by the Trades Union Federation in September 1990 that forced the regime of General Denis Sassou Nguesso to permit the formation of multiple parties. In Zaire, democratic forces had been agitating for change for years, and Mobutu's April 1990 concessions had been preceded by widespread political unrest, including major demonstrations in Kinshasa. The multiparty reforms President Paul Biya conceded in Cameroon in 1990 and 1991 were largely a response to civil unrest, including student protests and a series of demonstrations and riots during April 1991 that saw a reported 100 demonstrators killed.[29] Domestic opposition forces were demanding change even in repressive Equatorial Guinea, where there were numerous arrests throughout the early 1990s and violent anti-government demonstrations in 1993.

International pressures, however, were also a critical force in bringing about political change, however limited. Members of the US Congress had launched a debate over repression in Zaire, and over US policy toward Mobutu, as long ago as the 1970s. Restrictions on US economic and military aid to Zaire were first imposed by Congress in the mid-1980s; and by 1990, the executive branch was also demanding democratization. Belgium and France, allied with the United States in a 'troika' demanding change in Zaire, also applied critical pressure. The Belgians were particularly angered by the brutal suppression of student unrest at Lubumbashi University, (where an estimated 50 to 150 died in May 1990), and the ensuing cover-up. This incident led to a cut-off in Belgian aid and contributed to subsequent US and French decisions to suspend assistance. The French decision to move the 1991 francophone summit from Kinshasa to Paris, after Belgium and Canada threatened a boycott,[30] was a particular blow to Mobutu's image. At the June 1990 Franco-African summit, held at La Baule, France, President François Mitterrand had urged democratic change on the francophones generally, explicitly linking aid and democratization. Mitterrand warned that 'French aid will be lukewarm towards authoritarian regimes and more enthusiastic for those initiating a democratic transition.'[31]

Yet, while Central Africa's authoritarian leaders seemed to be yielding to domestic and international pressures for human rights and democracy reforms in the early 1990s, by mid-decade the reform movement had stalled nearly everywhere. Reforms appeared to succeed for a time in Central African Republic, where 1993 legislative and presidential elections were judged generally free and fair by international observers. However, the elected president, Ange-Félix Patassé, was soon being criticized as an ineffectual leader, beginning to show authoritarian tendencies, who was unable to turn around the country's economic decline. A military mutiny broke out in May 1996 over pay and living conditions, and violent clashes continued into 1997 despite efforts by France and African peacekeepers to restore peace.

Congo (Brazzaville) appeared for a time to be preparing successfully for elections but internal faction fighting resumed in 1997, casting the prospects for successful democratization into doubt. Limited progress in democratic reform was made in Gabon, where the principal opposition parties concluded the 'Paris Accords' with parties backing President

Bongo in October 1994. In July 1995, Gabonese voters endorsed legal and constitutional changes embodying the agreement in a referendum regarded as free and fair. The Accords were intended to assure inclusion of opposition figures in government, respect for human rights, and appointment of an independent National Electoral Commission.[32] Opposition parties, however, alleged massive fraud when legislative elections, which had been postponed several times, were finally held at the end of 1996.

After sidestepping French pressures for democratic change in Chad for years,[33] Idris Déby was elected president of that impoverished country in controversial elections in July 1996. But according to the Department of State's 1996 human rights report, the vote was marked by 'widespread and credible reports of fraud, vote rigging, and irregularities'.[34] In most of Central Africa, the stalled democratization drive is reflected in post-poned elections, blatant irregularities in the elections that have been held, and continuing press restrictions as well as other human rights violations that restrict democratic participation in the political process.

Democratization has stalled in Central Africa partly because the opposition in most countries has been fragmented along ethnic lines or because of the personal ambitions of rival leaders, making it possible for the region's autocrats to evade genuine democratization even while making surface concessions to democratic forms. Opposition parties fielded seven candidates for the presidency in Cameroon in 1992, but if they had united behind a single candidate they could almost certainly have ousted Biya, who received less than 40 percent of the vote. No doubt there were irregularities in this election, as opposition candidate John Fru Ndi and the US-based National Democratic Institute alleged, but if the opposition had been united these irregularities probably would not have been sufficient to keep Biya in power. And in Gabon, 12 candidates ran against President Bongo in December 1993.

Another factor delaying political change in Central Africa has been the surprising strength and resourcefulness shown by the region's authoritarian rulers in the face of domestic and international pressures. Until Mobutu's May 1997 overthrow, they had compiled a remarkable record as masters of delay and political schemers of Machiavellian caliber. Mobutu beat a strategic retreat on many issues, even losing control of the Central Bank, but after seeming to open the way to democratization in April 1990, he retained the presidency through seven years of cat and mouse games with his domestic and foreign critics. Most analysts felt that Mobutu would arrange matters so that he would prevail when the oft-delayed elections he had promised were finally held. Mobutu's fate was sealed only when his Rwanda-backed enemies in the east discovered that the ill-paid Zaire army would not fight.

In other countries in the region, however, opposition leaders who have risked imprisonment or even death have not yet proven a match for the region's authoritarian regimes. Equatorial Guinea's Teodoro Obiang Nguema, for example, while not as skilled as other regional leaders in political maneuverings, has successfully relied on the heavy hand of repression to prevent political change. Arrests, detentions, and beatings have repeatedly been employed against the opposition. Nonetheless, the opposition remains active,[35] and there is some potential for political change even in this country.

The future of democratization in Central Africa now hinges to a significant degree on the course of events in Congo (Kinshasa), which, as the regional giant, could be a major force for change, both as an example and a potential leader. Kabila promised elections in 1999, following the drafting of a new constitution and a possible constitutional referendum. The world greeted these promises with some skepticism, particularly in view of Kabila's ban on political parties and his highly strained relations with Tshisekedi. Kabila appears to be strongly influenced by President Museveni's view that in the African context, multiple political parties inevitably line up along ethnic lines and exacerbate ethnic tensions. But should democracy – even a limited, no-party democracy – take root in the huge Democratic Republic of Congo, authoritarian regimes in Gabon and elsewhere will face added difficulties in holding on to power.

Such regimes might still find some limited support from France, however. On 18 July 1996, President Chirac gave a speech in Brazzaville before both chambers of the Congo parliament which implied that his approach to democratization in Africa differed from Mitterrand's, as enunciated at La Baule. While Chirac spoke glowingly of the advantages of democracy and insisted that Africa should be done with *coups*, juntas, and violent transitions, he criticized conditionality in aid and lending as a means of bringing about reforms on grounds that it leads to a 'démocratie de façade'.

Instead of conditionality, Chirac maintained, France would promote reforms through inspiration and encouragement, recognizing that every country follows its own way to democracy.[36] Coming in the wake of a resumption of French contact with Mobutu, and an April 1996 French pledge to resume aid to Zaire, the Brazzaville speech raised questions about future French cooperation with the Western donors in pressing for democratic reforms. France continued to ally itself with Mobutu as the Zaire crisis deepened, and in November 1996, Chirac said of Mobutu, 'It is quite obvious that the international community cannot ignore that he is the man best placed to represent Zaire and find a solution to this problem.'[37]

France's influence in Central Africa, however, has declined significantly in the wake of Mobutu's ouster, and its future policies with respect to democratization may have less impact than in the past. Moreover, the election of a socialist National Assembly in France in June 1997 has reduced Chirac's ability to set the course of Africa policy.

Economic Trends

Every country in Central Africa has undertaken some sort of economic liberalization program aimed at reducing the government role in the economy and strengthening free-market forces. Typical measures have included privatization of state-owned enterprises, curbing controls over prices and exchange rates, reducing budget deficits, and various measures to promote foreign trade and investment.

Normally such measures could make the region more attractive to US and other Western investors, and lead to increases in trade as well. But in every Central African country, economic reforms have been half-hearted at best, and the IMF, the World Bank, and the bilateral donors have repeatedly had to apply heavy pressure to keep governments from abandoning reforms altogether. Governments have repeatedly sought delays in undertaking potentially unpopular measures, such as cutting back bureaucracies. They well recognize that reducing the state's economic role means that power and perquisites are also reduced. Consequently, reforms in Central Africa have not gone as far as in Côte d'Ivoire, South Africa, Ghana, or some other countries where political leaders appear more convinced of their necessity. GDP growth rates in the mid-1990s were negative in several countries, with strife-torn Burundi and Rwanda recording the worst performances. Only oil-rich Equatorial Guinea recorded economic growth rates significantly exceeding population growth.

The IMF has approved a new Extended Structural Adjustment Facility (ESAF) for Congo (Brazzaville),[38] which is privatizing the state water and electricity companies, telecommunications, and the agency responsible for rail and river transport as well as ports.[39] Expectations are generally low, however, in view of Congo's past performance. One diplomat has observed, 'They have been producing 150,000 barrels [of petroleum] a day for 10 years, and there is nothing to show for it but a $4 billion debt.'[40] Cameroon's reforms, particularly improvements in public finance management, have won approval from the Paris Club of donors in the form of a rescheduling of 50 percent of the country's public debt. The Club also promised to re-examine the entire question of Cameroon's debt stock within three years if reforms continue.[41] The IMF has approved a three-year series of loans to support Chad's economic adjustment program,[42] but whether Chad can recover from years of turmoil and mismanagement is unclear. Some economic activity has been renewed in Rwanda, but it too will need years to recover from conflict in 1994. Burundi, now under regional sanctions in the aftermath of the July 1996 *coup*, has not yet reached the bottom of its long, steep slide.

In Zaire, before Kabila's take-over, Prime Minister Kengo wa Dondo had achieved some success in reducing inflation and sparking the interest of foreign investors through a modest economic liberalization program. Kabila's economic policies remain to be clarified, although he has encouraged investments by foreign mining companies that have pledged to revive the industry. On the other hand, in the late stages of the civil war, he seized control of the railways in the east and south from a South African-Belgian consortium that had made great strides in rebuilding rail services in the region.

The former Zaire labors under a $14 billion foreign debt, including penalties for unmet arrearages, accumulated during the Mobutu era. About $1.5 million is owed to the United States, according to US Treasury data, while Belgium and France are also major creditors. Kabila has not repudiated the foreign debt, since doing so would mean most new aid and lending would be cut off. Privately, however, Kabila supporters argue that the Western governments and international financial institutions never should have lent such vast sums to the Mobutu regime, which was widely known to be corrupt. Congo has little to show today for these past borrowings, they argue. Lenders and the Kabila regime are consulting on easing this debt burden, and the success of their efforts will have a major impact on the former Zaire's economic future.

Regional Prospects

In each country of Central Africa, political and economic problems are so severe as to make pessimistic scenarios for the region's future entirely plausible. The break-up of the Democratic Republic of Congo, which could well have occurred in the early 1960s but for the United Nations intervention, is again a possibility. Shaba province is more closely linked to the economies of southern Africa than it is to Congo's, and there are

Lunda political leaders with no fondness for Kabila who might take up the secessionist cause as a route to personal power and wealth. The Kasai region, with its diamond wealth, a vibrant Roman Catholic Church, and Luba ethnic population is also functioning with a high degree of autonomy from the rest of Congo, and this region too was a focus of secessionist sentiment in the early 1960s. Observers judge that it will not soon attempt a formal separation, partly because of the fear of revenge against the widely dispersed Luba elsewhere in Zaire,[43] but if events should take an ugly turn in the conflict between Tshisekedi and Kabila, East Kasai's course could prove unpredictable.

Chad, with its armed rebel bands and clan-based politics, bears a certain resemblance to Somalia before its collapse; and a new Sahelian drought, which is an ever-present possibility, could sorely strain the political system – as drought did in Somalia. The destructive rounds of militia violence in Congo (Brazzaville) could recur indefinitely, since no one side seems capable of a decisive victory.

In Rwanda and Burundi, it is difficult at the moment to foresee a political settlement that will allow the Hutu and the Tutsi to live together in peace. Each group is highly sensitized to the threat from the other, and minor incidents can readily escalate into widespread violence. A large-scale international peacekeeping force, which might impose order in this region, will almost certainly not be deployed, and there appears to be little possibility at present of separating the Hutu and Tutsi into separate national states. Consequently, further violence, probably spilling across state boundaries, must be anticipated. The risks of political collapse are probably less severe in other Central African countries, although each has its own sharp social divisions and longstanding political grievances.

While pessimistic outcomes cannot be excluded, there is some chance that the momentous changes in Zaire in 1997, together with the inherent attraction of the region's resources, could contribute to a more positive future – one marked by economic growth running ahead of population increase, gradual improvements in respect for human rights, and increased observance of democratic standards in the conduct of political life. South Africa could prove to be a major contributor to this more positive future.

The end of South Africa's economic isolation, after democratic elections in April 1994, is already benefitting the wider region. White-ruled South Africa had long had a covert economic role in Central Africa, but today, South Africa's firms are free to operate in the open, and they are taking advantage of regional opportunities. Mining, railways, air transport, and tourism are areas in which South African companies hold considerable strengths, and their proximity as well as their experience in the African business climate could give them a comparative advantage over US business in the region. Rail links between the mineral-rich regions of Congo (Kinshasa) and South Africa already exist, and could help to forge a formidable economic alliance between the two countries. Nelson Mandela has cultivated good relations with Kabila and sent a top-level team to Kinshasa to assess prospects for economic cooperation.

South Africa's Afrikaner farmers are showing interest in Central Africa because of the plentiful rainfall. Rich soils found in many tropical areas hold out the prospect of profits in the production and export of tropical products. If growth in South Africa itself accelerates to the 6 percent to 8 percent level some anticipate by the end of the century, its economy will be a powerful force indeed, creating an expansion of trade through much of sub-Saharan Africa. Remittances sent home by expatriate workers in South Africa, many of them illegal immigrants, will make their own contribution.

Implementation of the Angolan peace accord, which continues despite the upheaval in Zaire, could also contribute to regional economic growth by promoting increased trade. The railway across Angola to the port of Benguela was the traditional route for transporting Shaba's minerals – as well as Zambia's copper – to the sea. Damage to the railway is extensive, but if it should be restored, regional transport will be greatly facilitated.

Positive scenarios for Central Africa are of course by no means assured, and any likely scenario can only be weakly positive, at least initially, in view of the serious problems the region confronts. Nonetheless, it would be premature to write off Central Africa as a region without prospects or to argue that it will soon be entirely marginalized in world affairs.

US Policy

Until the overthrow of Mobutu, post-Cold War era Central Africa, for reasons suggested in the introduction to this chapter, was not a key focus for US Africa policy. Stalled political change and halting economic reforms dictated that US policymakers, who place democracy promotion, sustainable development, and trade expansion at the top of the Africa policy agenda, direct most of their attention to other regions and countries. The impact of Central Africa's problems on US policy is suggested by Table 3.4, which reports the Administration's request to Congress for assistance to the region in FY 1998. Prepared before

Table 3.4   Fiscal 1998 administration aid request ($ millions)

	Development	Emergency Food	Peace Corps	Military Education
Burundi	2.0			
Cameroon			2.8	0.1
CAR				0.1
Chad			1.0	0
Congo			1.1	0.2
Eq. Guinea				
Gabon			2.4	
Rwanda	7.5	44.8		0.3
Zaire				

Source: US Agency for International Development congressional presentation tables.

Mobutu's ouster, these data underscore the very limited expectations US policymakers had for the region.

Only Rwanda was targeted for a significant amount of US assistance, and most of the program there would be devoted to emergency food aid aimed at facilitating the resettlement of returned refugees. Other aid was aimed not at promoting economic reform, reflecting US disappointment with reform programs in the region, but at easing the lot of the poorest of the poor. Most of this aid is going to health-related projects to reduce the incidence of HIV/AIDS, foster family planning, and cut the rate of child mortality. The Peace Corps is working directly with the poor in small-scale programs similar to those found elsewhere in Africa. The small military education programs in the region are democracy-oriented, reflecting a very modest effort to increase professionalism and respect for democratic values among officers. Clearly, US policymakers have not believed that significant objectives can be realized in Central Africa under prevailing political conditions.

Environmental projects in Central Africa, particularly its rainforests, were slated for increased support in future years in keeping with environmental interests among the American people and many policymakers. The US Agency for International Development plans to launch a $14 million, five-year project to reduce deforestation, encourage conservation, and promote biodiversity in the tropical forests of the Congo Basin.

The absence of planned aid for Zaire, often a leading recipient of US aid in sub-Saharan Africa, was not a reflection of the absence of a policy, but rather part of the effort to push Mobutu toward democratic reforms. Economic aid was suspended; there was no training or other aid for the

Zairian military; and visas were not provided to Mobutu's associates or to anyone else who had profited from the corrupt political system in Zaire. Prime Minister Kengo, whom Washington regarded as a pragmatist and a technocrat, was, however, being encouraged in the reforms he had undertaken; and in March 1996, he spent three days in the United States on what was described as a private visit but included meetings with National Security Advisor Anthony Lake and other US officials.

Meanwhile, the United States sent an ambassador, Daniel Simpson, an experienced diplomat, back to Zaire to try to encourage the electoral process, primarily through continuous meetings and discussions with Zairians across the political spectrum, including President Mobutu. Policymakers noted that apart from the presidential vote, the legislative elections might be highly important if they created a parliament whose members had some sense of accountability to the voters. Provincial elections were also seen as critical in creating more responsive government in Zaire.

The changes in Zaire could well lead to a revival of US assistance in the region, and to a much higher level of diplomatic activity. In 1997, as Kabila's troops advanced without encountering significant impediments, the US Ambassador to the United Nations, Bill Richardson, spearheaded an urgent round of negotiations in the region aimed at arranging what he called a 'soft-landing for Mr. Kabila's rebels when they reach Kinshasa, a landing that avoids bloodshed and chaos'.[44] A month later, Richardson returned to Congo and reportedly discussed a three-year, $50 million US aid program for the Democratic Republic of Congo.[45] US Agency for International Development sources have not confirmed this report, and while Congo has received some $10 million in emergency, non-food humanitarian assistance in 1997, a more substantial aid program is still under study. Richardson, referring to the alleged atrocities against Hutu during the war, has noted what could become an important condition on US aid, telling Kabila, 'We want reports of those massacres investigated; we want those involved in those massacres punished.'[46] Congress and the Administration will find it difficult to approve a large new aid program for the Democratic Republic of Congo unless investigations of the alleged atrocities are carried out and a satisfactory explanation, exonerating Kabila, is found. Nor will aid be approved in large amounts if Kabila fails to implement his promised democratization program.

In short, large-scale US engagement in promoting the recovery and rehabilitation of Congo (Kinshasa) is quite possible but by no means assured. If it does come about, US interest in the region as a whole could increase as the changes in Congo promote an overall strengthening of Central Africa's economies and political systems.

Overall, the United States has deployed limited policy instruments in Central Africa in the post-Cold War era, with limited expectations. The recent overthrow of Mobutu may now have removed the central obstacle to a better future for Central Africa and its people, and if this proves to be the case, US involvement in the region will increase. Should Kabila's Congo fail to overcome the Mobutu heritage, and fall into political and economic difficulties, major policy initiatives will continue to be deferred.

Notes

1. This figure comes from the US Department of State's *Country Reports on Human Rights Practices for 1994*, Washington, DC (1995), 201. Higher estimates are often reported.
2. US Department of State. Remarks by George E. Moose (4 April 1995) Assistant Secretary of State for African Affairs to the Annual Trade and Investment Conference on Africa, San Francisco, California.
3. See, for example, William D. Scott (1993) *Called to Africa* (Little Rock: August House); and Peter J. Brashler (1979) *Change, My Thirty-Five Years in Africa* (Wheaton, IL: Tyndale House).
4. US Library of Congress (1994) Federal Research Division, *Zaire: A Country Study* (4th edn), 39–42.
5. Fred E. Wagoner (1980) *Dragon Rouge: The Rescue of Hostages in the Congo* (Washington, DC: National Defense University, Research Directorate). See also, Thomas P. Odom (1988) *Dragon Operations: Hostage Rescues in the Congo, 1964–1965, Leavenworth Papers*, no. 14.
6. *Facts on File Yearbook* (New York: Facts on File, Inc., 1967), 309.
7. *Facts on File Yearbook* (New York: Facts on File, Inc., 1977), 265.
8. US Library of Congress, Congressional Research Service, *Zaire – Political Transition: Issues for the United States. CRS Issue Brief* 92033 by Raymond W. Copson (archived).
9. Reuters report (14 November 1996).
10. Raymond W. Copson (1994) *Africa's Wars and Prospects for Peace* (Armonk, New York: M. E. Sharpe), 127.
11. US Library of Congress, Congressional Research Service, *Rwanda and Burundi: US Policy Options. CRS Issue Brief* 94027 by Raymond W. Copson (archived).
12. 30 June 1994 press release.
13. Hearing (22 July 1994).
14. These data include economic as well as military assistance, although military assistance programs in the region during the period covered were not large, consisting mainly of small military education and training grants.
15. The 'Brooke Amendment', Sec. 512 of the Foreign Operations Appropriations (Public Law 104–208), prohibits most aid to countries more than one year in default on loan repayments. Disaster assistance and some other types of aid are exempt.

16. US Department of the Interior, Bureau of Mines (January 1995) *Mineral Commodity Summaries, 1995* (Washington, DC), 47.
17. 'Equatorial Guinea on the Brink of an Oil Boom'. Reuters report (25 February 1996).
18. 'Congo Sets Up Fund to Manage Debt, New Oil Cash.' Reuters report (24 January 1996). See also, 'Congo-Brazzaville Becomes a Boom Town', *Africa Economic Digest* (22 January 1996).
19. *Africa South of the Sahara, 1996* (London: Europa); 286.
20. Howard W. French (3 April 1996) 'An African Forest Harbors Vast Wealth and Peril', *New York Times.*
21. 'Sudan's War Threatens Zaire Wildlife with Extinction', Reuters report (14 February 1996). Reuters was citing the World Wide Fund for Nature.
22. US Central Intelligence Agency, *World Factbook, 1995.* In Gabon, however, the rate of population increase has fallen below 2 percent.
23. Saskia Van Hoyweghen (July 1996) 'The Disintegration of the Catholic Church of Rwanda', *African Affairs,* 379–401.
24. Reuters report (3 October 1994).
25. John Pomfret, 'Rwandans Led Revolt in Congo'.
26. Thomas Turner (1994) 'Government and Politics', in Sandra W. Meditz and Tim Merrill (eds), *Zaire: A Country Study* (Washington, DC.: Library of Congress, Federal Research Division), 263.
27. *Africa South of the Sahara* (1996), 311.
28. *Africa Research Bulletin* (December 1990), 9939.
29. *Africa South of the Sahara* (1995), 231.
30. *Africa Research Bulletin* (January 1991), 9713–14.
31. Guy Martin (March 1995) 'Continuity and Change in Franco-African Relations', *Journal of Modern African Studies,* 15.
32. Department of State (1995) Country Reports on Human Rights Practices.
33. 'Paris donne six mois à Idriss Déby', *Jeune Afrique* (16–22 November 1995), 38–9.
34. Department of State (1995) Country Reports on Human Rights Practices, 55.
35. *Country Reports on Human Rights Practices, 1995.*
36. Text provided by Embassy of France. For a discussion of the speech, see François Soudan, 'France-Afrique: les mots et les choses', *Jeune Afrique* (7–13 August 1996), 36–7.
37. Reuters report (5 November 1996).
38. 'IMF to Vote on New Accord for Congo', Reuters report (13 March 1996).
39. 'Congo Sacks 122 Anti-Privatization Strikers', Reuters report (25 January 1996).
40. Howard French, 'Sleepy Congo, a Poor Land Once Very Rich', *New York Times* (18 June 1995).
41. *Africa Research Bulletin* (16 November 1995 to 15 December 1995) 12380–1. Economic Series.
42. *Africa Research Bulletin* (16 August 1995 to 15 September 1995) Economic Series, 12683.
43. 'Kasaï Takes Off', *Africa Confidential* (19 January 1996).
44. Reuters report (5 May 1997).
45. *Washington Post* (8 June 1997).
46. *New York Times* (7 June 1997).

4

US Policy Towards East Africa: Crisis Response amid Limited Interests

James E. Winkates

For purposes of this chapter, East Africa is defined broadly to accommod-ate a wide geographic area of the land portion of the African continent and the Indian Ocean archipelagos. Coverage embraces the Sudan, Eri-trea, Ethiopia, Djibouti, Somalia, Uganda, Rwanda, Burundi, Kenya, Tanzania, and Madagascar; brief mention will be made of the Indian Ocean island groups of the Seychelles, Comoros, Mauritius, and the French Department of Reunion. Perhaps more than any other African region, East Africa exhibits very wide physical, cultural, economic, and political diversity. It is the thesis of this chapter that US interests in the wider East African region beginning with the pre-independence period, during the Cold War, currently, and in the longer-term future have been and will be limited; furthermore, that these interests will be treated as intermittent and constrained in the larger context of other US global interests and the perceived requirements of domestic politics. That said, nevertheless, there exist opportunities in the region for the US to advance its national interests and, at the same time, to make a construct-ive contribution to the needs of the states in the area.

Given geographic distance, US interests historically have been quite limited. Unlike the special relationship with Liberia in West Africa, confrontational experience with the Barbary pirates in North Africa, Cold War intrigue in Zaire, and a proxy civil war in Angola, US interests have been intermittent and often derivative of other international or domestic concerns. As the lone remaining superpower, however, US interests necessarily remain broad, if not always deep, even in those world regions that are not central to national destiny.

Before examining official US interests and policy in this region, a quick panoramic view of popular concern is in order since visual images of East Africa regularly appear in US media. Presently, astronomists point to a 15-billion-year-old universe. The earliest archaeological finds of hominids by the renowned Leakey family in Kenya and Ethiopia have been chronicled repeatedly on American television and in popular magazines. Electronic and print outlets regularly feature the Nile River, the world's longest, and the rich fauna of the Ngorongoro Crater. The 1990 trade ban on elephant ivory and subsequent Convention on International Trade in Endangered Species, which outlines protection for rhino and gorilla among other region-specific wildlife,[1] are known to elementary school children. The East African ethnic diversity and cultural richness have increasingly invaded popular American culture. Manute Bol, a seven foot six inch young man from the Dinka tribe in the Sudan, became the tallest player in the history of the professional National Basketball Association.[2] The most photographed tribal group on the continent, the Masai, frequently grace national newspapers, magazines, and video media. Despite genocidal carnage in Rwanda and Burundi, the Tutsi tribe are depicted as statuesque warriors. The Makondes of Tanzania, artisans and sculptors, have acquired a world-class reputation among museums and individual art collectors. Coastal Arabs plying their dhows greet viewers on the *Discovery* and *Learning* channels on a regular basis. Even the stories of remote pastoralists, the Karamojong of northern Uganda and Kenya, and the ancient religious community of Falasha Jews, airlifted to Israel in two humanitarian waves from their Ethiopian homeland, have made their mark in national news media. Two movie classics, *The African Queen* and *Khartoum*, often replayed for late-night viewing audiences, have now been joined by more recently produced movies such as *Out of Africa* (Kenya) and *Gorillas in the Mist* (Ruwenzori Mountains of Rwanda). Region-specific Africana, pictorial and print, have punctuated the popular American culture. In the future the diversity of positive popular interests may cultivate more public support for US official policies in the region.

Pre-Modern US Contacts

Throughout the isolationist phase of American history, which continued until World War II, the United States largely ignored the African continent. Most of the major interior explorations of East Africa originated in London (Speke, Burton, and Livingstone). European and American slave ships plied the West African coast, not East Africa's. US commercial interests, other than the slave trade, were negligible. No US military forays took place in the region. An early and obvious anomaly, the US inaugurated economic contact with East Africa through a treaty of commerce and amity with Zanzibar in 1833. The pact provided for most-favored-nation trade status and fixed minimum tariff charges.[3] Nevertheless, American interests in Africa generally, and East Africa specifically, were minimal, essentially reflected in missionary activity, modest and specific trade in raw materials and cash crops, and private philanthropic contacts. Insularity and preoccupation with building a nation to meet the industrial age governed American thought. Although invited to participate in the partition of Africa at the 1884–5 Berlin Conference, the US attended, but declined any offer of territory. President Cleveland, nevertheless, exacted assurances of humanitarian treatment for Africans in this colonial scramble and, with unaltruistic vision, the guarantee of an open door for US commerce. England, France, Germany, Belgium, and Italy all took their spoils, building on the Portuguese presence that had been on the continent since the days of exploration with Vasco da Gama. Not until the acquisition of the Philippines in 1898 did the US establish a permanent overseas presence.

Following World War I, at the Versailles Peace Conference, Africa attracted only minor attention. Former German territories (including present-day Tanzania, Rwanda, and Burundi) were given to colonial European powers and South Africa to manage as League of Nations mandates. Despite President Woodrow Wilson's moral penchant for self-determination and as a principal architect of the mandate system, the US voiced no special concern for East Africa. As in Berlin earlier, the US declined any responsibility for mandated territories. Failure to ratify the Versailles Treaty and join the League effectively surrendered any US influence of colonial policy or practice in the region.

The only measurable US presence was the foreign missionary contingent in the region, which continued to grow larger and more diversified throughout the twentieth century. Protestant overseas missions dominated the field. Denominationally, American Methodist, Presbyterian, Seventh-Day Adventist, and Pentecostal groups were the most significant, Africa-wide and in East Africa.[4] There have been two other groups of note. The Sudan Interior Mission (renamed more recently the Society for International Ministries or SIM), an evangelical, interdenominational, Protestant group operating in the Sahelian belt and into the northern rim of East Africa, has maintained a presence in East Africa since 1927. More recently, Jehovah's Witnesses have been especially active in Kenya. Except for Tanzania, Catholic missions in the area have been staffed mostly by clergy from France, Belgium, and Italy. Many, if not most, of the nationalist leaders who brought independence to East African countries were products of mission schools.[5] To this extent, private US church programs indirectly advanced independence movements in the region.

Mussolini's forcible takeover of Ethiopia stirred American sympathies. Emperor Haile Selassie's plea for help against Italian aggression before the League of Nations proved poignant but ineffective in prompting much political, let alone, military response. In fact, as Italian forces moved to take over Ethiopia in August, 1935, Congress passed the Neutrality Resolution, which imposed a mandatory embargo on arms and ammunition exports to all belligerents upon the outbreak of war. That occurred in October 1935. The League made a half-hearted, unsuccessful effort to invoke economic sanctions against Italy. On the contrary, US exports to Italy, including oil, rose during the early months of the Italo-Ethiopian war.[6] Like western European governments, there proved to be insufficient wrath to reverse the occupation until Allied forces evicted the Italians in 1941. During World War II Secretary of State Cordell Hull pleaded for the colonial powers to 'fix, at the earliest practicable moment, dates upon which the colonial peoples would be accorded the status of full independence'.[7] The US overtly supported time limits for the independence of Italian colonies, including a ten-year window for Italian Somaliland. At the close of the war, the US lead in founding the United Nations provided ample opportunity to press the issue of colonial self-determination. However, resumption of partisan domestic political debate and rival efforts to restore the pre-war international system engulfed senior policymakers.

The emergence of the Cold War virtually ensured US deference to colonial control in Africa. Nationalist sentiment in Asia, notably India, Malaya, and Indonesia, peaked soon after the onset of the Cold War. The United States showed no interest in challenging European powers about self-determination issues in sub-Saharan Africa, as they were already burdened with independence struggles in Asia. With the creation of NATO in 1949, Britain, France, Portugal, and Italy soon became linchpins in maintaining Western solidarity in the face of confrontation with the Soviet-Warsaw Pact.

Thereafter, two key events caused great concern in Western capitals. In 1955, Egyptian President Gamal Nasser nationalized the Suez Canal, prompting British, French, and Israeli forces to invade and recover control of the waterway. The US opposed its allies in this venture and, ultimately, succeeded in getting the UN to provide a face-saving way out for the invading forces. Also in 1955, the then non-aligned nations met in Bandung, Indonesia, to form a bloc of neutral states that rhetorically objected to capitalistic and communistic political ideologies, but began to vie with East and West for economic and military aid monies. This bidding war for political affinity and UN General Assembly votes later evolved into a near-automatic crisis-response strategy in Washington and Moscow. Symptomatic of official US caution to avert further Third World instability, Secretary of State John Foster Dulles chose to emphasize the 'orderly development of self-government' to manage the gradual dissolution of post-war European colonial empires.[8] Contrarily, reflecting a minority view, senior US diplomat Chester Bowles publicly proclaimed the lack of a positive African foreign policy, contending that 'the most powerful country in the world ... cannot declare itself to be a nonparticipant in the affairs of a continent boiling with change, without abdicating its position of leadership'.[9] Nevertheless, the US Cold War priority of political stability in contested areas translated into going slow in encouraging early independence movements.

Before 1957, US satisfaction with the political status quo and with colonial European management of East African affairs, as well as preoccupation both with world crises and domestic priorities, America remained relatively indifferent to the region. Only feudal Ethiopia was independent at the time. Sudan became the first colony in the region to achieve its independence (1956). When the Gold Coast gained independence from Britain in 1957 as Ghana, the rush of decolonization began for all of sub-Saharan Africa. Upon his return from the Ghanaian independence celebration, Vice-President Nixon recommended the creation of a new Bureau of African Affairs within the State Department. Congress authorized the new bureau in 1958. At this time there were more Foreign Service officers assigned to West Germany than in all of the African continent.[10]

Throughout the post-World War II era, the US remained philosophically uncomfortable with the continued colonial dominance of the African continent, in part because its own civil rights revolution had been accelerating. President Truman had issued an executive order in 1949 to integrate the armed forces. In 1954, the Supreme Court unanimously ruled in *Brown* vs. *Board of Education of Topeka* that 'separate was no longer equal'. In the legislative arena, the 1964 and 1965 Civil Rights Acts largely closed the legal loopholes to segregated housing, eating, and entertainment facilities. As would later be the case as well, US interest in sub-Saharan Africa was derivative of domestic politics and the rising tide of the civil rights movement.

Perhaps more importantly, growing US interest also stemmed from the shattering of colonial rule in black Africa. The historic tension in US foreign policy, between the values of order and justice, by 1960 had begun to tip toward justice. The eruption of the Mau-Mau rebellion in Kenya (1952), the onset of the bloody Algerian civil war (1954), sporadic anticolonial rioting in several African capitals, and the Sharpeville massacre in South Africa (1960) provided impetus for a changed policy environment. Finally, British Prime Minister Macmillan's 1960 'Winds of Change' speech in South Africa encapsulated a new era. The inauguration in 1961 of a Democratic US President provided the rationale for a new look at emerging Third World nations. The policy question became not whether to support African independence struggles but rather how.

US Relations with East Africa since 1960

The political transition from President Eisenhower to President John Kennedy resulted in greater attention paid to Africa generally. The first diplomatic appointment of Kennedy was former Michigan Governor G. Mennen ('Soapy') Williams to be Assistant Secretary of State for African Affairs. The Kennedy circle of policy advisors reflected only a modest substantive difference from the Eisenhower Administration on African affairs, but the policy tone and emphasis changed greatly. Freshly independent African nations were embraced as new and hopeful players on the international scene. Kennedy immediately established US embassies in every new capital. As East African nations gained their self-determination, the US often provided economic aid and technical assistance. Major philanthropic foundations, such as Ford, Rockefeller, Carnegie, and the Phelps-Stokes Fund, inaugurated many new grant projects in agricultural development, education, and site-based training to aid the fledgling governments. Many African students came to study in the US through the beneficence of foundations, universities, and the federal government. With the exception of Nigeria, Ethiopia sent more students to the US than any other sub-Saharan nation.

Once the torrent of independence movements began, colonial submission proved surprisingly swift and, with some exceptions, largely a peaceful process. In sequence, the first wave to independence in East Africa included Sudan (1956), Somalia and Madagascar (1960), Tanzania (1961), Uganda, Rwanda, and Burundi (1962), and Kenya (1963). The second wave, the Indian Ocean archipelagos, started with Mauritius (1968), followed by the Comoros (1975) and the Seychelles (1976). Finally, after 30 years of civil war, Eritrea (1993) completed the process. The pattern for all of Africa, including this region, was twofold. Early independence came first to those areas in which the indigenous, protonationalist movements proved most assertive. Secondly, there was no economic rationale for early takers or late arrivals. For example, the politically astute Julius Nyerere took Tanzania, the least economically developed of the British East African colonies, to independence in 1961, a year ahead of Uganda, and two years ahead of Kenya, the most economically advanced of the trio. However, the real turning point for African independence occurred in 1960. Seventeen African states joined the United Nations that year. These were heady times for Africa and the United States, which for the first time had to develop foreign policies (political, economic, military, and cultural) for 25 percent of the UN voting membership.

Three major influences impacted nascent US policy toward Eastern Africa. These geopolitical factors were cast more in the form of general principles rather than region- or country-specific policies. First, the Kennedy Administration pursued a policy of denial toward all leftist-leaning African political movements, a posture designed to thwart growing Soviet and, to a lesser extent, Chinese political influence. Second, the rapid expansion of African membership in the UN necessitated US efforts to seek allies in the ideological combat of the Cold War. Third, the new policy elite found a moral imperative to aid the emerging nations and at least attempt to help them achieve their first steps toward stability and development. A basket of assistance programs resulted. These included official development grants and loans, administered through the Agency for International Development (AID), Export–Import Bank Loans, humanitarian Food for Peace shipments, and a vigorous Peace Corps presence in the region. In addition, official encouragement to the private sector expanded bilateral trade with the region.

Since East Africa had not become a crisis area for US policy nor one where any vital nor important interests were at stake, official aid disbursements started and remained quite modest. Rupert Emerson nicely summarized the aid and trade profile of the region within the larger African context for the post-World War II period.[11] As he pointed out, the distinguishing characteristics of official US aid were its modest proportions and the small military component. US aid was intended only to supplement the assistance given by the former colonial powers. For the entire period 1948–65, only two East African states received even modest US assistance.

During this period the Mahgreb countries, Zaire and Nigeria led the aid list. Ethiopia, never colonized, had no aid patron beyond modest Italian contributions in Eritrea. For a limited time, only Ethiopia and Sudan received more than negligible US aid monies. US military assistance for the same 1948–65 period for all of Africa totaled $186.1 million of a

Table 4.1   Assistance monies provided by AID and its predecessor agencies ($)

Country	Loans	Grants	Total	FY 1965 Only Total	Food for Peace Cumulative FY 1946–65
Ethiopia	36.5 m	68.2 m	104.7 m	7.9 m	14.7 m
Sudan	15.8 m	55.3 m	71.1 m	2.7 m	18.4 m

Source: R. Emerson. *Africa and United States Policy* (Englewood Cliffs, NJ: Prentice-Hall, 1967), Table 3, p. 185.

$34.7 billion global total. Ethiopia received $95.6 million cumulatively, the largest US recipient on the continent. Although still modest in relative dollars, reasons for the aid should be noted. The only remaining Department of Defense facility in East Africa, the communications and listening post at Kagnew Air Station in Eritrea that the US had operated under a 1953 base agreement, justified the bulk of the expenditures. In addition, Addis Ababa had no colonial patron, maintained a stable, pro-Western government, and by 1963 had become the site for the head-quarters of the Organization of African Unity (OAU). By 1961–2, the Congo crisis consumed the bulk of policymakers' time and energy to the detriment of East Africa.

In the early 1960s, similarly, US trade with all of sub-Saharan Africa barely exceeded $2 billion, a third of which year in and year out was with South Africa. (At that time total US world trade amounted to $45 billion). For FY 1964 US exports to East Africa totaled $50 million and imports $165 million. Exports typically included machinery, autos, chemicals, and textiles. Imports were mined ores and raw agricultural products, especially coffee from Ethiopia, Kenya, and Uganda, hides from Ethiopia and Somalia, and spices from the island archipelagos. However, the region's resources and purchasing capability were only marginally important to the US economy. Economist Andrew Kamarck's judgment, that the elimination of imports from Africa and African markets for American exports would cause only an imperceptible ripple in our standard of living, still holds true.[12] For animal hides, processed meat, coffee, tea, and bananas upon which East Africa is dependent for hard currency, there exist many alternative and more accessible market sources for the US.

Emerson further recounts the minimal US private investment in the continent during the mid-1960s. While the dollar value had increased rapidly, it paled in contrast with European investments and comprised but a small fraction of total US world investment. US African investment more than doubled in the decade of the 1950s, but it still amounted to less than 3 percent of total US foreign investment by 1960, rising to $1 billion of a total $35 billion globally.[13] Sub-Saharan Africa accounted for the least amount of US private investment in all of the Third World. In general East African nations received negligible investments because the region had few major extractive industries and no petroleum production.

In the heady days of the JFK 'New Frontier' the creation of the Peace Corps greatly impacted East Africa. Volunteers to Africa nearly equaled Latin America's as the largest geographic contingent. By 1965, Ethiopia hosted the second largest group in sub-Saharan Africa, and nearly all continental states in the region hosted Peace Corps volunteers. The flow of East African students studying in the US was no less notable. In 1965, four of the ten African states sending more than 25 students to the US annually were East African (Kenya, Tanzania, Ethiopia, and Uganda).[14] At this time, federal and foundation monies flowed freely to support international education, especially area studies and African area emphases. Grants for faculty and student exchange programs, art exhibits, dance troupes, conferences, and speaker programs linked even small communities and liberal arts colleges with people and things African. These early educational thrusts were illustrative of a US-African 'window of opportunity' not seen since.

The Eclipse of US–East African Relations, or 'Things Fall Apart'

The decades of the 1970s and 1980s saw a significant downturn in US relations with the region, again both for intra-African reasons as well as causes external to the area. Certainly, the grip of South-East Asia on US attention and resources colored all else. On the continent, the extended Congo crisis, the Biafran civil war, and the implacable apartheid problem occupied American attention to the near exclusion of all else. Put simply, Washington sought politically viable and economically friendly states with which to conduct normal relations. US relations a priori assumed the character of denial strategies to limit all communist penetration and involvement in the region. The Cold War had overlaid the continent.

Key events in the East African region virtually ensured a growing estrangement in US relations until the early 1990s. Military *coups* and mutinies enveloped many states, beginning with a military take over in the Sudan (1958, and again in 1969), an attempted *coup* in Ethiopia (1960), mutinies in Kenya, Uganda, and Tanzania (1964), and *coups* in Zanzibar (1964), Burundi (1966 twice, 1976), Somalia (1969), Uganda

(1971, 1980), Rwanda (1973), and Ethiopia (1974). Aborted *coups* also occurred in the Seychelles (1981) and Kenya (1982). Repeated political destabilization by nascent militaries proved to be a common phenomenon in post-independent East Africa.[15] Persistent interventionism by East African militaries, a pattern indeed reflected throughout the continent, gave major cause for the US not to invest too fully its energies in the region until some degree of normalcy returned to the scene.

Washington officially encouraged African military support of UN peacekeeping operations, most especially that of the Ethiopian military, which had been trained by US forces and had performed well in Korea and later in the difficult Congo operation.[16] Ironically, the roles of African militaries and the modest US efforts to keep channels open with military governments proved divisive within Congress. Oftentimes, Congressional liberals were more inclined to look with sympathy on military take overs while conservative representatives maintained strident opposition to all *coups*, sometimes with racial overtones.[17] US African policies in this period often derived from, or were influenced considerably by, evolving domestic US political ideologies.

Compounding US relations were several diplomatic incidents and terrorizing threats to US personnel and citizenry in East Africa. In 1965, two US diplomats accused of ill-defined subversive activities were expelled from Tanzania. Again, under peculiar circumstances, Burundi expelled the American ambassador in 1966.[18] Increasingly, Cold War issues defined US policy in the region. In the wake of the October 1969 military *coup* in Somalia, the US ended its economic assistance program because Somali flag vessels continued to trade with North Korea and Cuba.[19] Subsequently, the Soviet Union became Somalia's largest military and bilateral economic aid donor.

Two countries, Sudan and Uganda, well illustrated erratic US relations in the region. Following the 1958 *coup*, all Sudanese political parties were banned. Even upon resumption of the political process, the army intervened repeatedly during the period 1965–89. Except for the 1972–83 period, Sudan has been engaged in civil war since 1963, with the formation of the Anya-nya guerrilla movement in the southern provinces. During the Congolese civil war, in which the US sought desperately to find and support a central government that had credibility, Sudan allowed the trans-shipment of arms from Algeria to the Simba rebels in northern Congo.[20] However, the US severed relations in 1967, because of Sudan's support of Egypt and the Soviet bloc against Israel. In a bloodless *coup* in 1969, Colonel Nimeri again put the military in charge, and again relations were discontinued. Without opposition, he was elected in 1971.[21] An attempted *coup* in July 1971, led by the Sudan Communist Party, resulted in a limited *rapprochement* with the US. Shortly after the Black September terrorist attack on Israeli athletes at the Munich Olympics, that same terrorist group in March 1973 kidnapped and murdered US Ambassador Cleo Noel, his deputy George Moore, and a senior Belgian diplomat at the Saudi embassy in Khartoum, an early precedent toward the internationalization of terror against diplomats, which the host nation has responsibility to protect under international law.

In 1977, Nimeri expelled Russian diplomats but also overtly supported Somalia in its claim to Ethiopian territory. By 1978, the Carter Administration made the Nimeri government the principal regional beneficiary of US aid, largely to counter Soviet influence in Ethiopia and to gain support for US policy in the Middle East. Following the assassination of Anwar Sadat of Egypt, in 1981 the Reagan Administration again substantially increased military aid to Sudan. But Nimeri's economic mismanagement, including the doubling of external debt, the imposition of economic austerity measures required by the IMF, and an inability to defuse internal conflict, led to a *coup* that overturned him in 1985.

The erratic behavior of Sudan continues to date. Alternating postures on the application of Islamic law, the Sharia, resulted at one point in 58 public limb amputations for convicted thieves, crucifixion sentences, and death for converting from Islam to Christianity.[22] For much of this period the US withheld all but humanitarian aid. All other aid was terminated after the most recent *coup* in 1989, in which General Omar Hassan al-Bashir emerged as head of the Revolutionary Command Council for National Salvation, later called the National Islamic Front. In 1993 the Department of State added Sudan to the shortlist of state sponsors of terrorism. The Khartoum government for a time protected the infamous international terrorist, Carlos the 'Jackal,' and provided safe haven for the Palestinian Hamas and Hizbollah terrorist groups. Relatedly, five Sudanese suspects had been arrested in a New York City bomb plot.[23] And three Sudanese nationals are suspected in the attempted assassination of Egyptian President Hosni Mubarak while in Addis Ababa in 1995. In early 1996 the UN Security Council debated imposition of sanctions against Sudan unless the government extradites the suspects in that attempt for trial.

Of greatest concern to the US presently is the potential for Sudan to incite Islamic extremism in the region. In the North African neighborhood, violent confrontation between Islamic moderates and militants in

Algeria has already cost 75,000 lives and is rising weekly. Sudan's close association with Libya and its not-so-covert support for various terrorist groups, including active training camps, make for a volatile base for extremism in the region. Both Eritrean and Ethiopian officials have expressed concern that Sudan may be infiltrating Muslim communities within their borders, perhaps through returning nationals displaced by the lengthy Ethiopian Civil War.[24] Islamic fundamentalism has spread to predominately Christian Uganda, so far without major incident. Arab monies have been used to establish an Islamic university, to fund lower-level educational institutions, to launch banks, and to start several newspapers.[25] Sudan's eastern and southerly neighbors confront an expanding fundamentalist agenda in states that have so far averted religious war. In February 1996 Washington ordered the first 25 of its remaining diplomatic corps in Khartoum to depart the country because of 'the continuing concern for the safety of American officials' and suggested that Americans living there 'may wish to consider departing at this time' as well.[26] By late March 1996, all US diplomatic personnel had departed. At the same time, in the first election since he took power in a 1989 *coup*, Lt.-Gen. Omar el-Bashir gained 75 percent of the vote. Hassan Turabi, the fundamentalist Muslim cleric widely believed to be the prime mover behind the government, also won a seat in the parliamentary elections.[27] The state remains largely isolated from normal relations with most African, Middle Eastern, and Western governments.

Similarly, modest US relations with Uganda plummeted swiftly, but for different reasons. A former British colony, at independence Uganda consisted of an amalgam of monarchical fiefdoms and assorted, competitive, and vengeful ethnic groups. The first prime minister, Milton Obote, a Langi tribesman, abolished the traditional kingdoms and took the new state sharply toward Marxist socialism, taking care to elevate his own tribesmen to key positions in the bureaucracy and the military. Armed with a socialist economic master plan and a nation endowed with considerable human and agricultural advantages, ethnic favoritism on a grand scale shattered the socialist vision.[28] The army revolted and Idi Amin, Chief of the Army and a Kakwa tribesman from the remote north-western region, assumed power in January 1971. Until forced out by Tanzanian military forces in 1979, the bizarre Amin presided over and participated actively in the brutal torture and massacre of countless of his fellow countrymen.

Two examples that prompted deteriorating relations will suffice. Amin's security organs (army, police, state research bureau, and the secret police) were neither instructed nor motivated to provide safety for domestic or foreign personnel in country; in fact, they were the greatest danger to civic peace. Amin publicly condoned the actions of Palestinian extremists, such as the Black September murder of US diplomats in Sudan, and he praised publicly the actions of Adolph Hitler in eliminating minorities in Germany. In October 1973 the US chargé d'affaires was summoned to meet Amin's new foreign minister, an army colonel, who relayed Amin's dictum that all Marine security guards were to be expelled within 48 hours for alleged subversive activities.[29] Although no official reason was given, in all probability it was Amin's anger at the rapid US military resupply of Israel in the October war. The US had few investments, official or private, in Uganda and the loss of diplomatic presence was of little direct consequence to America. The largest contingent of Americans were missionaries dotted in posts throughout the country. The Peace Corps program had been shut down two years earlier, and the small economic aid program had been terminated in midsummer 1973. The parting words of the US chargé, as the last diplomat to leave Uganda, were poignant. He noted that 'the United States can get along without Uganda' and that 'the tragedy is what one man, who is unfit for the office he holds . . . has done, is doing and will continue to do to ruin this country of Uganda and its 10 million people'.[30]

The second episode occurred on 27 June 1976, when four members of the Popular Front for the Liberation of Palestine (PFLP) terrorist group hijacked an Air France commercial flight with 257 passengers and crew *en route* from Tel Aviv to Paris and forced the plane to land at Entebbe, Uganda. Ultimately, all but Jewish hostages were released by the PFLP during negotiations. Amin and his security forces clearly were supportive of the hijacking and aided the terrorists on the ground at Entebbe. The US put its intelligence community assets, especially the CIA and FBI, in support of the Israeli plans for a rescue. In a dramatically successful raid, an Israeli military force retrieved all but four hostages safely on US Independence Day, 1976. On at least four counts US interests were well served with the Israeli success. Innocent victims were saved, an ally's moral mission was aided, a brutal dictator was humbled, and international terrorists were defeated. Noteworthy was the reaction of African governments and the press, succinctly put – 'public anger, private approval'.[31] The rescue fortified the application of US moral principles and invocation of the eleventh commandment, 'Thou shalt not bow down to terrorism.'

When the Tanzanian forces overturned Amin and entered Kampala in 1979, neither the US nor the world in general voiced opposition.

Amin fled to Libya, and later to Saudi Arabia. Former Prime Minister Obote returned through rigged elections, which saw the resumption of intertribal strife at a cost of 300,000 lives and 1 million refugees.[32] In 1986, Yoweri Museveni led his National Resistance Army, armed by Libya, and deposed Obote for a second time. The US remains hopeful, even somewhat optimistic, that President Museveni may ultimately succeed in restoring order and sustained economic progress to Uganda.[33]

In both instances, Sudan and Uganda in the 1970s, US official and private interests were indeed minimal and policy often derived from interests external to the country in question and US policy therein. Neither situation related directly to US policy toward that government. These cases illustrate that where US interests are not vital, nor even of major importance as is the case in East Africa, US policy becomes reactive not pro-active, and cannot be otherwise. Since 1996, the US maintains a normal diplomatic presence in Uganda, but only a Sudan Interests Section in our Nairobi embassy. Academic retrospects on US policies in Third World nations readily confirm that much of the stimulus behind US policy had little to do with Third World governments, but rather with other US global concerns.[34] That judgment certainly applies to US policy toward East Africa.

Cold War Sideshows in East Africa

While the major Cold War battleground in Africa had been the former Belgian Congo, gaining its bloody independence in 1960, both the US and the UN paid a heavy price for the successful effort to stabilize that government. The aftertaste for US policy in other areas, such as East Africa, resulted in only modest engagement. In the wake of the 1974 revolution in Ethiopia, clearly the key US client state in the region, Emperor Hailie Selassie was deposed and later murdered. A military government took over, soon led by Col. Mengistu Haile Mariam, an army officer and head of the ruling military council known as the Dergue. His radical Marxist regime invited in approximately 1,000 Soviet advisors and 11,000 Cuban troops, at their peak, to train and supply arms to his military forces, engaged in guerrilla war both against Ethiopian and Eritrean rebels in the north and Somali rebel forces in the east. The US soon withdrew completely from Kagnew Station in Asmara, relying for a time on US installations in Turkey and the emergent capabilities of satellite technology for monitoring communications traffic in the Soviet Union and the Middle East. Today, that former installation is a rehabilitative center for the many war amputees who served in the long Ethiopian civil war.[35] To be sure, the USSR proved increasingly active in the East African region with arms transfer agreements in the 1960s and 1970s. Moscow delivered arms to Sudan, Somalia, Ethiopia, Tanzania, Kenya, and Uganda. At the same time, all of these states were sequential or simultaneous recipients of Western military assistance, all of which much preferred Western sources, if available, because of their better quality, reliability, and access to spare parts. Egypt's defeat in the 1967 Arab–Israeli War, relying heavily on Soviet military equipment, sent a strong negative signal to Moscow's arms clients. Only Moscow's arms pipeline to Ethiopia lasted the decade of the 1980s.

The People's Republic of China (PRC) agreed to be the principal donor in the construction of the Tanzam railway, running from Tanzania to Zambia, to alleviate Lusaka's reliance on Rhodesian railways for its access to an Indian Ocean port in South Africa. Zambia depended on Rhodesian rails for a full three-quarters of its imports in the early 1970s. The Salisbury government, having declared its Unilateral Declaration of Independence (UDI) from Britain in 1965, virtually controlled Zambia's economy. Begun in 1968, this massive project required an investment of $450 million.[36] A large contingent of Chinese workers and their political commissars injected another Cold War factor onto the East African landscape. At this point the Chinese had become the continent's number one aid donor, offering twice Moscow's amount of economic aid. For Peking it was the competition with the Soviet Union, not with the United States, on which they focused in an effort to emerge as the true ideological patron in the African sphere. Then Premier Chou En-lai used the Tanzam railway initiative as a demonstration project to secure a leadership position in the Third World. Soon the Great Cultural Revolution, however, turned China severely inward until nothing came of Chou En-lai's further great hopes.

In the wake of the Iranian revolution and the Soviet invasion of Afghanistan in 1979, the proximity of East Africa's Horn to the Persian Gulf increased the strategic importance of Somalia, Sudan, and Kenya to provide closer military access, if needed, to execute the Carter Doctrine, which declared protection of Middle East petroleum flows a vital US interest. Proximity of the Greater Horn region to 70 percent of the world's petroleum reserves prompted perceived threats of an 'arc of crisis' stretching from Afghanistan to the Horn. Relatedly, the prospect of Soviet interdiction of the oil tanker sea lanes fortified policy convictions that the US had to help secure the Horn and the sea lines of communication from the Bab el Mandeb strait to the Red Sea and the Suez Canal. Both claims seduced US policymakers but proved more mythical than real.[37] These myths and delusions were vastly overstated by various US domestic ideological constituencies such as the Committee on the Present Danger in the 1980 presidential election campaign. Bilateral economic and military assistance to Horn countries grew dramatically in a very short time, peaking in 1984–5, and then plummeting just as quickly. Indeed, during the Ford, Carter, and early Reagan Administrations, geopolitical Cold War considerations directed US Horn region priorities.[38]

Perhaps the most resonating of Cold War intrigues in the region occurred in 1977. Until then, Somalia had been a Soviet client on the Horn. Indeed, in 1972 the Soviets had been granted a modest naval anchorage and air access at Berbera, on the Gulf of Aden and in close proximity to the Bab el-Mandeb strait, southern entry to the Red Sea. While the Soviets had managed to maintain client relationships with both Ethiopia and Somalia, the Somalis claimed much of the Ogaden, a vast eastern region of Ethiopia, mostly populated by ethnic Somalis. Skirmishes between Ethiopian troops and Somali rebels occurred regularly. In 1977 Ethiopia accused Somalia of sending large numbers of troops and heavy weapons into the region. Russian military forces were expelled from Somalia in retaliation for what appeared to be Soviet support of the Ethiopian territorial claim in the Ogaden War. In 1978, some 11,000 Cuban troops armed with Soviet equipment, defeated Somali troops and ethnic rebels sequestered in Ethiopia. Ironically, the Soviets were forced to choose between the two governments, leaving it open for the US later to come to Somalia's aid. The Soviets did acquire use of a Red Sea military base on the Dahlak islands in 1979, in return for $1.5 billion in military arms to the Ethiopian army and double that figure in military aid between 1979 and 1986.[39] Those gains, however, never translated into significant benefit for Moscow.

At the time, the US Government regularly voiced great concern over the large presence of Soviet and Cuban troops on the Horn. In retrospect, these scenarios proved only sideshows for both superpowers. The key tests of staying power for both Washington and Moscow had been a long way off in the jungles of Vietnam and the mountains of Afghanistan, where both superpowers saw their hopes of controlling Third World political-military conflicts dashed.

Certainly with 20/20 hindsight it is easy to conclude that the superpower competition for Cold War favor in East Africa proved only an ephemeral phase yielding little long-term benefit either to the major powers or to their diverse African clients. The migration away from a policy prism of 'benign neglect' in the immediate post-World War II decades to an intermittent and pulsing arms competition changed little in East Africa. However, the Cold War greatly exacerbated active civil conflicts like that in the Sudan and dormant interethnic hostility more recently evident in places like Somalia. For Moscow's last East African client, Ethiopia, the militarization of that multi-ethnic country ultimately brought down the government in 1991 and resulted in the emergence of the region's newest independent state, Eritrea, depriving Ethiopia of any direct access to the Red Sea.[40] Whatever lessons may have been learned by the US in this Cold War interlude have proven of little value in the post-Cold War security environment of the 1990s, where intrastate breakdown, not external aggression, predominates.

Somalia 1992: A Catastrophic Case

President Mohammed Siad Barre, vanquished earlier in his quest to reunite all Somali people in the Ogaden with Somalia, lost whatever claim he had to political legitimacy in 1991. His government, based on a coalition of clans with blood ties to him, proved unable to counter rebellion in Somalia's north, a movement encouraged by the similarly failing Mengistu of Ethiopia. Barre fled Mogadishu, with the combative clan groups controlling pockets of territory, but lacking the necessary political consensus to establish an effective, united government. By 1992, thousands of Somalis were dying of starvation weekly, and international relief agencies put out a desperate call for immediate food and medical supplies. In April 1992, the UN authorized the United Nations Operation in Somalia (UNISOM I), whose purpose was to provide humanitarian aid and to facilitate the termination of hostilities in Somalia, an initiative fully supported by the US Government. In August 1992, President Bush ordered a massive airlift of food to arrest the worsening and widespread starvation. In September 1992, Andrew Natsios, the president's special coordinator for Somali Relief, testified that unless clan conflicts could be ended, there was no hope to move from a dire state of emergency to long-term rehabilitation for Somalia.[41] Throughout that fall, approximately two-thirds of the relief supplies were stolen or sold on the black market by armed militiamen called technicals. Looters attacked food convoys and the aircraft coming in to drop supplies. The multinational force on the ground could not secure routes of passage nor distribute the desperately needed food supplies. Consequently, in December 1992, President Bush committed the first group of US ground forces, growing quickly to 28,000 troops. They occupied Mogadishu airport to create a more secure environment for the humanitarian relief operations. To get food to outlying areas, the troops accompanied relief trucks to regional distribution centers. The massive airlift effort became the most extensive US commitment to the Horn of Africa, and to the region, to date.

When in May 1993 the multinational operation escalated to UNISOM II, which embraced extensive nation-building efforts, President Clinton was asked to provide a Quick Reaction Force to conduct military operations to safeguard further UN efforts at political reconciliation as well as continuation of the major relief effort. Both the UN effort and the key support provided by the US confronted an irreconcilable situation. Food had become a weapon in clan combat. To ensure the initial policy of emergency food relief could succeed, a more activist use of ground troops to deter and to disarm the roving militias proved necessary. This state of affairs resulted in bloody skirmishes and later confrontations with local Somali, and especially with the armed protectors of the local warlords. In the most dramatic confrontation and firefight, 18 US soldiers were killed in ambush and some of their bodies dragged through the streets of Mogadishu for the worldwide television audiences to view. That was the last straw. The President announced that all US forces would leave Somalia by March 1994, at which point the entire operation was turned over to UN multinational peacekeepers. They, too, left in March 1995. To ensure their safe retreat, 1,500 armed US soldiers and 350 Italian marines kept the warlord militiamen out of the port of Mogadishu and then withdrew. The World Food Program, UN Children's Fund, and the UN Development Program remained to combat disease, chlorinate wells, and repair roads, schools, and hospitals. The International Red Cross, Doctors Without Borders, and the US-based, private, World Vision International maintained missions in Somalia. The UN mediator and his staff are headquartered in Nairobi and meet with the leading warlords only in Kenya. Little has changed politically, but there is no longer outright starvation. Approximately 350,000 Somalis died in the year before the US ground forces insertion, most because of starvation.[42] Another drought or cholera epidemic could easily amplify that bleak obituary.

The entire Somali operation cost the US approximately $1.2 billion, besides a UN outlay of $1.5 billion. Forty-four US soldiers were killed, and 175 wounded. 140 UN peacekeepers were killed. Many thousands of Somalis died as a result of interclan violence.[43] Much has been written on the multidimensionality of the Somali tragedy.[44] From the perspective of

US interests and policy, there were compelling humanitarian needs, but no vital US interests were at stake. Further, the utter frustration with events in Bosnia at the time, where Serbs tortured, raped, and murdered Moslems in officially sanctioned 'ethnic cleansing', in part impelled US policy to respond to what appeared to be a more manageable humanitarian rescue operation in a culturally homogeneous, all-Muslim state. The coalition relief effort involved 20 UN member states and scores of relief organizations. These were good faith efforts, undertaken with the best of intentions though coupled with much naïveté, but in the end establishing a very negative policy precedent for future, large-scale, US commitment toward the Horn, specifically, and to the miseries of East Africa generally.

Not incidentally, the insolubility of Somalia's problems terminated the bloom on 'New World Order' hopes of a more benign, perhaps more manageable, post-Cold War environment. The stakes were less, yet the dangers in many ways more unpredictable and intractable. The lack of any tolerance for US casualties in the sparse landscape of Somalia pointedly illustrates the low level of US conviction that it could make a difference, or that it ultimately mattered. As Chairman of the Joint Chiefs of Staff, General Shalikashvili admitted he could not conceive of anyone intervening in Somalia again. He concluded, 'They are on their own.'[45] More than any recent failed initiative, US policy in Somalia also resulted in a congressional tightening in the funding and support of UN operations of a peacekeeping nature; moreover, the intermittent resurrection of the neo-isolationist impulse in the 1996 Presidential campaign owed much to the Somalia experience.

Ripple Effects of Genocide in Rwanda, then Burundi

Interethnic confrontation between the majority Hutu and the minority Tutsi in Rwanda and Burundi has been a recurring, tragic saga with both ethnic groups responsible for periodic brutal massacres. In 1994 genocide erupted when the plane carrying both presidents home from peace talks mysteriously exploded. Certainly the US emergency response was first and foremost a humanitarian effort, but beyond that the massacre of 500,000 Rwandans, Hutu and Tutsi alike, accelerated mass migration of refugees into neighboring states. In Congressional testimony in mid-1994, Brian Atwood, AID Administrator, drove home this very point.[46] He recounted that of Rwanda's 8 million population, 5 million were refugees or internally displaced. Approximately 1.5 million fled to Zaire, 460,000 to Tanzania, 200,000 to Burundi, and 10,000 to Uganda, states that are incapable, and largely unwilling in any event, to care for this cross-border deluge. At the regional level, USAID further estimated that 23 million people are at risk of starvation or displacement due to drought, ethnic strife, or chronic food shortages in the Sudan, Ethiopia, Eritrea, Djibouti, Somalia, Rwanda, Burundi, Kenya, Uganda, and Tanzania.

To avert jeopardizing modest political and economic gains among Rwanda's neighbors or seeing further retrogrades, Atwood pointedly discussed a refugee containment strategy. To support the multilateral UN relief effort for Rwanda, the US deployed a joint task force with 4,000 troops. Food, medical supplies, water purification systems, oral rehydration packets, blankets, and shelter materials were flown in on round-the-clock schedules. The immediate goal, to safeguard the health and well-being of the refugees, expanded to the creation of safe zones in Rwanda so that ultimately the refugees would return home and no longer threaten very fragile neighboring state infrastructures.

Rwanda, not unlike Somalia, swiftly became another 'failed state', without a workable government, enmeshed in torture and retributions, and spilling millions of its citizens quickly into bordering states. In conjunction with the UN Aid Mission in Rwanda (UNAMIR), authorized by the Security Council and the UN High Commissioner for Refugees (UNHCR), only the US possessed the long-range air transport capability, airport construction kits, and relief supplies to surge quickly and prevent an unparalleled human disaster. To be sure the relief operation involved many nations and UN agencies; nevertheless, it is the US that gets the call first in these large-scale disasters. Not surprisingly, Rwanda's genocide spilled over into neighboring Burundi, which shares the same proportional ethnic composition as Rwanda. In both cases, approximately 85 percent of the population is Hutu, but the governments, militaries, and economies are largely controlled by members of the minority Tutsi ethnic group.

By mid-1996, 150,000 Burundians had been killed in an ethnically inspired civil war. Burundi's third president, Sylvestre Ntibantunganya, a Hutu in a government dominated by Tutsis, tried to lead a national reconciliation. He failed. In July 1996, he fled to the US embassy because of uncontrolled violence in the capital. The Tutsi took power immediately in a military *coup*. The ambush and apparent slaying by Hutu rebels of Catholic Archbishop Joachim Ruhuna, a Tutsi, in September 1996, pushed the small nation to the brink of chaos.

By October, 1996 in Arusha, Tanzania Secretary of State Warren Christopher raised the prospect of yet another humanitarian intervention anticipating further deterioration of civility in Burundi. In particular, he had tried to persuade neighboring states to establish an African Crisis

Response Force, that would be trained and funded by the US.[47] Again, the US found itself in a crisis response demand, trying to fashion a coalition of states in the region that might serve to confine violence with the US as paymaster.

While Washington may no longer be the lone '911' number to curb major military conflict as in the Cold War period, it is the primary emergency number for mega-disasters that require immediate, large-scale response to contain what can easily become even graver catastrophes. The point being the number of 'failed' or 'failing' states in East Africa alone (Somalia, Rwanda, Burundi and potentially Ethiopia) can only suggest a kind of containment and triage strategy for which there is no US stomach but neither an easy moral retreat.

Post-Cold War Assessment of US Interests and Policy

Dissolution of the USSR in December 1991, the resulting tumultuous and halting moves toward democracy and market capitalism in Eastern Europe, combined with large US budget deficits and political clarion calls for progress toward balanced budgets, further marginalized US policy in East Africa. The most succinct official summary of US interests in Africa generally that also applies to East Africa appears in the *National Security Strategy*, a document published near annually by the White House. The current edition avers the promotion of market democracies, respect for human rights, sustainable economic development, and resolution of conflicts through negotiation, diplomacy, and when necessary emergency peacekeeping. Acknowledging a degree of 'Afropessimism', the policy statement notes nevertheless that Vice-President Gore has made two extensive trips to the continent and recounts the first ever White House Conference on Africa in June 1994. At the same time, the policy tone is clear. The document asserts that 'efforts by the United States and the international community must be limited in duration and designed to give the peoples of a nation the opportunity to put their own house in order. *In the final analysis, the responsibility for the fate of a nation rests with its own people*' (emphasis added).[48] The basic policy thrust is that the US will work with international organizations (UN, World Bank, IMF), regional agencies (OAU and the African Development Bank), and non-governmental organizations (NGOs), e.g. Red Cross, Doctors Without Borders, CARE, on the urgent issues of population growth, health and disease, environmental degradation, debt relief, trade and investment promotion, among others, but the primary action responsibilities devolve on African governments.

What has changed dramatically since the end of the Cold War for the West, and the US certainly, is the criteria by which Western governments are willing to engage in sub-Saharan Africa on their terms. Thomas Callaghy, for example, made an early telling case that the post-1989 political changes in the former USSR, conjoined with Africa's politico-strategic marginalization already widely perceived even before the Cold War's end, meant that both favorable economic and political conditions must exist if Congressional aid is to flow.[49] This dual conditionality is evident in stated US interests to encourage market democracies and sustainable economic development. East Africa, to be sure, has few assets (strategic importance, geographic proximity, or scarce resources) to warrant special treatment.

On a normative level, one may reasonably ask if US policy is 'moral' or 'right' in the imposition of conditions for further US involvement. Furthermore, are those conditions 'fair' and 'realistic'? The answers to both questions are a resolute 'yes'. Participatory government by those who are governed, sustainable development, and preservation of human rights together go a long way to ensure indigenous civic cultures operating to the benefit of the nation. Moreover, such performance can often negate both external intervention and various dependencies that can permanently erode state sovereignty. On the other hand, one can easily argue that both scarcity of official time and monies may be 'immorally' and 'unfairly' spent on questionable international recipients to the detriment of other, more deserving, candidates or national interests of higher priority. Lastly, the record to date also suggests that in mega-disasters the US contributes a healthy share towards international rescue operations.

Two recent assessments of US global strategic interests have a bearing here. The Institute for National Strategic Studies of National Defense University, Washington, DC, appraised US interests and key concerns for the African region as a whole, with some application to East Africa.[50] In summary, it found US national interest 'minimal' militating against significant involvement, but also noted a policy quandary. Namely, the US wants the international community to respond to crises but seeks no US lead nor major responsibility. The study found three East African governments illustrated positive development signs: Eritrea, Uganda, and Mauritius. In all three instances, state political and economic performance conform to stated US policy interests, or at least there is positive movement in that direction.

Eritrea, victor in a 30-year civil war with Ethiopia and with a meager population of 3.5 million, exhibits what appears to be a squeaky-clean national government under President Afworke, who has made it clear this generation must sacrifice for its children and grandchildren.[51] Uganda, under newly elected President Museveni, has been stabilized politically and has embarked on economic reforms, processes which have been rewarded with a $30 million USAID development grant. Mauritius, diversifying its private sector economy away from a near-total reliance on sugar production to textile exports, is fast becoming a minor regional economic 'tiger'.

The second recent assessment is one from the conservative think-tank, The Heritage Foundation.[52] This 1994 study found no vital US interests nor threats present in East Africa. Inferentially, if the Bab el-Mandeb Strait or Red Sea were blocked from oil tanker transiting, if terrorism would re-erupt in Sudan or elsewhere in the region, or if Islamic extremism were to gain a foothold in several countries, only then might important or vital interests come into play. The shortlist of threats in the region include human rights abuses and poverty that might destabilize an entire region. However, the report notes the region's positive trends in moving toward more open political and economic systems, privatization, and the propensity towards freer elections. The Foundation's assessment concludes there remains a need for targeted, but limited, US developmental aid and humanitarian relief.

While much has been made of the 'clash of civilizations' argument[53] to promote heightened concern over Islamic extremism, at least for East Africa it strains credulity. To be sure, the Islamic faith is spreading rapidly in the region, perhaps for quite mundane reasons. Islamic institutions provide a wide range of social services (food, shelter, health care) which governments have often little or no capacity to offer. Moreover, financial aid from the Arab Bank for Economic Development, the Islamic Development Bank, and the Islamic Solidarity Fund offer access to external funds not otherwise available. Interestingly, East African conflicts have been fundamentally ethnic, not religiously, inspired.[54] The Sunni Islamic sect predominates, and tends to adapt well to local cultures.

Bluntly, there are no vital security nor economic stakes for the US in East Africa. Unlike elsewhere on the African continent (Egypt in North Africa; Nigeria in West Africa; Zaire in Central Africa; and South Africa in southern Africa), no state in the region dominates politically, economically, militarily, or culturally. There is no likely hegemon in any plausible future scenario. It is the 'moral imperative' to which the US has responded when Sahelian drought, ethnic fratricide, and massive starvation occur. In 1993–4 the US spent over $3 billion on emergency peacekeeping and relief operations in but two countries, Somalia and Rwanda.[55] Understandably, if shortsightedly, the US Congress illustrates a distemper for most Third World aid, just as private relief agencies express widespread 'donor fatigue'.

Current and Projected US Interests and Policy

The four major components of national security policy (political, economic, military, and cultural) might best be used to assess longer-term US interests and preferred policy outcomes in the region. Certainly there is much commonality in policy prescriptions for East Africa as a whole, while at the same time specific US interests will determine sub-regional or even country-specific approaches. Perhaps the most significant feature of Cold War termination for the region is the message to all governments that they must lower their expectations for the substance and degree of US involvement in the area. Competing external and domestic priorities make the US much less sanguine about the prospects for resource allocation and policy prescriptions. The crux of policy choice becomes the relative gain anticipated for the resources expended. Foreign policy is resource driven, the two key resources being available monies and public support. Realistically, the US has the least to gain geo-economically in sub-Saharan Africa versus all other world regions. On the other hand, there can be political, economic, military, and cultural benefit from more US engagement as most states in the region would prefer more US interest and participation, especially in expanded trade.

Political Interests

The US maintains full diplomatic relations with ambassadorial status in all countries with the exception of Somalia (terminated 1994), Sudan (1996), and the Seychelles (1996). The US maintains a Sudan Interests Section in Nairobi. The embassy in Seychelles closed 30 September 1996. The decision to terminate the mission there stemmed from cost-saving elimination of small embassies and, more significantly, the 1995 decision by the Seychelles government to grant safe haven and immunity from criminal prosecution to anyone who invests $10 million in country.[56] That policy is an open invitation to drug lords. The US ambassador resident in Mauritius also represents US interests in the Comoros and in Diego Garcia, which is leased from the United Kingdom, and serves as the primary basing facility for US military forces in the Indian Ocean. This facility is a strategic, and arguably vital, US interest in South Asian waters and along with the US Fifth Fleet, based in Bahrain, major politico-military assets for timely logistic access where there are no US bases.

Politically, the US plays the role of catalyst and honest broker to nurture peaceful change, aid conflict resolution, promote stable demo-cratic pluralism, and sustain economic development. In 1991 when the Soviet Union imploded and abandoned its African interests, the US decided to stay engaged. When Mengistu fled Ethiopia at the conclusion of the civil war in 1991, the US mediated the talks that resulted in the peaceful resolution of the conflict and ultimately the Eritrean plebiscite for independence. In November 1993, the Organization for African Unity (OAU), headquartered in Addis Ababa, Ethiopia, established a conflict resolution and peacekeeping mechanism within its secretariat. The US helps provide the infrastructure support and shares American peacekeeping standards and experience. In 1994, President Clinton signed the African Conflict Resolution Act, which authorizes funding of sub-regional and non-governmental efforts at conflict resolution. The ongoing USAID 'Greater Horn of Africa' initiative creates a foundation for crisis prevention and food security. And, of course, the US is by far the largest donor to refugee aid and relief. Following the Rwandan mas-sacres, the US spent $400 million for humanitarian assistance to Rwanda and Burundi, and wrote off $1 billion in low-interest loans Africa-wide.[57]

Perhaps the most difficult political challenge in the region lies with Kenya, a country rent by deep tribal animosities, a culture of corruption, the highest birth rate on the continent, and a dismal human rights record. President Moi brooks no personal criticism. US pressure on Moi to migrate away from a one-party system led to Western aid donors freezing $350 million in earmarked aid in 1991. Only then did the President allow for new local, parliamentary, and presidential elections in late 1992. Thug attacks on opposition and human rights leaders, the muzzling of the press, and extrajudicial executions occur commonly. In 1994 Amnesty International reported scores of people killed by govern-ment supporters, frequent arrest, detention, and imprisonment of oppo-sition critics, and trumped up treason and sedition charges levied against political opponents.[58] A recent, feisty, US Ambassador, Smith Hempstone, granted asylum to a leading human rights lawyer, follow-ing a violent paramilitary police attack on political dissidents. Ambassa-dor Hempstone's account of Kenya's political machinations led him to question continuation of most-favored-nation economic status and to suggest draconian cuts in economic aid.[59] On both the political and economic fronts, Kenya may well have the potential to rival any state in the region if and when it gets its house in order.

Politically, the US promotes democratization, encourages regional self-reliance, and participates actively in multilateral approaches to chronic problems, but keeps the bilateral US effort minimal. US policy tries to stimulate, orchestrate, and prod states in the region to open up their political and economic systems, guarantee human rights, and develop 'civic cultures' that will enhance nation building. The key to accomplishment is securing access for the US political, economical, military, and cultural interests. With no *vital* regional interests, the current level of US presence and engagement assures maintaining sufficient access to further US policy objectives.

Economic Interests

What, then, is the US economic interest and attendant policy? Annual US contributions to the chief international loan sources for Africa, embodied in the African Development Bank ($9 million), African Development Fund ($131 million), and International Development Association, the World Bank's soft loan window ($835 million), total nearly $1 billion.[60] US direct investment for Africa as a whole approximates $5.5 billion (1994), the smallest private investment in any world region, or 1 percent of global US private investment. Half of that is in Egypt, South Africa, and Nigeria. Of the remaining $2.7 billion, the US Department of Commerce registers only $150 million in all of East Africa, including the island archipelagos, as follows: Kenya, $131 million; Mauritius, $8 million; Ethiopia, $5 million; Djibouti, $4 million; Rwanda, $2 million. All other East African states have only negligible US investment capital.[61]

Total US trade balance in the region is a modestly positive $51 million (1994). That breaks out by country (in millions) as follows in Table 4.2. Five East African states (Rwanda, Burundi, Seychelles, Comoros, and newly independent Eritrea) had insufficient US trade to register in official reporting.

One major consequence of Cold War termination has been a massive reduction in direct, bilateral foreign aid to the continent and into the region. Overall US aid to sub-Saharan Africa declined from $1.72 billion in the peak Cold War year of 1985, to $1.2 billion in 1992, and to a little over $800 million a year by 1993.[62] By 1996, foreign aid to sub-Saharan Africa dropped to $650 million. In October 1996, Secretary of State Warren Christopher made his first trip to the continent and into the East African region (Ethiopia and Tanzania) since a previous Secretary had done so in March 1990.[63] Foreign economic aid monitored and disbursed by AID for states in the region follows (in millions) in Table 4.3.

Table 4.2   US exports, imports, and merchandise trade
balance, by country: 1990–4

Country	Exports	Imports	Balance
Djibouti	7		7
Ethiopia	143	34	109
Kenya	170	109	61
Madagascar	48	57	−9
Mauritius	24	217	−193
Somalia	30	<500K	30
Sudan	55	35	19
Tanzania	49	15	34
Uganda	28	35	−7

Source: Table no. 1341, *Statistical Abstract of the United States
1995*, 819–22.

Table 4.3   US foreign aid-commitments for economic
assistance, by region and selected countries: 1985–93

Country	1985	1990	1993
Burundi	4	20	17
Kenya	40	36	16
Madagascar	5	17	44
Rwanda	19	12	6
Somalia	51	1	28
Sudan	149	8	–
Tanzania	<500K	9	23
Uganda	9	43	39

Source: Table no. 1333, *Statistical Abstract of the United States
1995*, 813.

Currently, the only two long-term AID projects in East Africa entail
$30 million to Uganda and $20 million to Tanzania. The US now targets
its very modest aid only to those nations undertaking democratic and
free market reforms, the twin tests of aid conditionality. In 1988, the US
played a major role in helping to establish the Multilateral Investment
Guaranty Agency (MIGA) as a World Bank affiliate, designed to stimulate direct capital investment flows to African states. To further encourage American corporate investment on the continent, in 1990 the US
initiated a $30 million Africa Growth Fund. So far, not surprisingly, the
US business response to both programs has been discouraging.[64] In 1994,
the US Government spent $1.1 billion helping US companies export
their products worldwide. That outlay translated into nearly $20 billion

in exports and approximately 300,000 American jobs.[65] Virtually none of those export sales went to East Africa.

Per capita gross domestic product (GDP) growth rates for the period 1961–87 showed a negligible or negative trend for six East African nations (Ethiopia, Madagascar, Somalia, Sudan, Tanzania, and Uganda), illustrating the extended (25+ years) negative baseline from which these states must emerge.[66] Of the fourteen East African countries included in this discussion, eight show in excess of 3 percent annual population growth, resulting in a predicted doubling of their populations in 22 years. While the US supports the rigorous IMF Structural Adjustment Programs in place in East Africa, it is nonetheless understandable given their historical economic performance how very difficult it is for central governments to service and reduce debt, cut government employment, privatize economies by terminating or selling off entrenched parastatal bodies (Kenya has nearly 150 still operating), and drastically curbing imports. These actions are a recipe for political defeat if not popular revolt. To partially alleviate some of the worst debt burdens in the world's poorest nations, the Group of Seven (G-7) in September 1996 agreed to relieve up to $7.7 billion in debts of as many as 20 of the most impoverished nations, most of which are located in sub-Saharan Africa.[67] The plan will cover up to 80 percent of the debt held by individual G-7 nations. No doubt several East African states will be among those selected.

For much of East Africa the US has overtly encouraged the further involvement of the more than one thousand UN-registered NGOs, private, voluntary, non-profit groups like the World Wildlife Fund for Nature, Doctors Without Borders, and CARE. Solutions to economic, social, and environmental problems often come more swiftly, and with significant local involvement, from NGOs than the public sector. These associations can transform into local, self-help groups, which can then tap into the resources and expertise of the parent association.[68] The US, in fact, might do even more in marrying up additional US NGOs with extant local union and craft associations, especially women's and children's rights groups, professional societies, agricultural cooperatives, and environmental and conservancy groups.

One good example of an international organization (state memberships) that has direct, practical relevance for East Africa is the World Tourism Organization (WTO). While Kenya has been the region's premier tourist destination, other generally stable governments (Tanzania, Uganda, Eritrea, Ethiopia, Mauritius, Seychelles, and Madagascar) could benefit from new or expanded links with the WTO. Certainly both advantages and disadvantages accompany the creation and sustainment of the tourist trade.[69] If planned well and regulated for local benefit, however, it can serve to supplement job creation, cultivate entrepreneurial support business, and bring needed hard currency to struggling governments. In the early 1970s, before the failure of the Selassie government, Ethiopia did a moderately good job in attracting foreign visitors on Ethiopian Airlines' 'Hidden Empire' tour of five historic sites. Ethiopian Airlines remains one of the few viable African carriers and indeed trains other continental carriers. US corporate business has much experience in planning and managing tourist facilities for a diversity of traveler requirements.

Security Interests

Currently, the US has no permanent, nor significant, military presence anywhere in East Africa nor Africa generally. There are no combat forces, no bases, and no homeporting of naval ships.[70] As with the political and economic components of US policy in the region, US presence and access remain the paramount objectives. US military programs support democratization and human rights, provide humanitarian assistance, encourage economic and social development, and attempt to nurture regional security arrangements. No state in East Africa confronts an external military threat from outside the region. In the wake of the Cold War, demands of the IMF Structural Adjustment Programs (now applied in 30 African states) and national budget reductions, all states are reducing the size of their militaries, some by as much as 50 percent. Moreover, the US and other Western aid donors increasingly demand military force reductions as criteria for further assistance.

The historic US relationship with Ethiopia (then including Eritrea), evidenced in the lengthy US military presence at Kagnew Station in Asmara, continues but on a smaller scale. At the request of those governments, the US advises on the retraining of the Eritrean and Ethiopian armies, both of which are engaged in significant military downsizing efforts. US personnel have also assisted in efforts to clear some of the many unexploded landmines which remain from the civil war. The Eritrean port facility at Massawa, which has six ship berths with off-loading capability, if modernized, could facilitate and expedite the import and export of goods and serve as a hub for humanitarian crisis response operations. The other Eritrean port at Assab is connected to the rest of the nation by an unpaved road.[71]

During the Gulf War and the Somalia relief operation, Kenya granted the US military use of the port and air field at Mombasa, which served as a staging area for trans-shipment of goods into the Gulf and Horn. The US maintained a satellite tracking station in the Seychelles, but that closed in August 1996 followed by the closure of the US embassy in September.[72] An alternative site will need to be found in the Greater Horn area or on a ship platform to ensure monitoring of US satellites.

US defense programs for sub-Saharan Africa generally fall into what has been called security assistance, renamed by President Clinton in 1994 as Assistance for Promoting Peace and Building Democracy.[73] Closer cooperation with African militaries is clearly desirable since they could be key players in the pursuit of political and economic reforms, guarantees of human rights, and further democratization. Two security assistance offices are located in the region, one in Djibouti and another in Kenya. In Asmara, a representative of the US Central Command (CENTCOM) serves as the regional liaison for the Tampa-based military command. CENTCOM maintains the chief responsibility for any US military forces sent into the Horn region. US embassies in Ethiopia and Madagascar host smaller defense attaché offices, which have responsibility for security assistance in their respective areas. The US now discourages conventional arms transfers to Africa. In 1985, the last peak year of US military aid into the region, only three countries benefited (Kenya, $22 million; Somalia, $34 million; and Sudan, $46 million) out of a then $279 million for all of Africa.[74] Both Somalia and Sudan were zeroed out in 1991 and Kenya in 1994. No East African government has received military loan or grant funds for the purchase of military weapons or equipment since 1994. Several small US military-funded programs aid states in the region. Madagascar has received money to construct low-cost housing, a military civic action project intended to fill a specific social need. Funding through the Military Health Affairs program permitted the purchase of infant incubators for Uganda.

Perhaps the most visible contribution is the International Military Education and Training Program. An average of 500 military officers and enlisted personnel from throughout Africa annually receive training or education at US military schools, a program that began in 1950. In 1991, this program expanded to include senior civilian officials in non-defense functions, legislative members responsible for the oversight of military programs, and more recently even non-government civilians, NGO personnel for example. These programs expose students to the American democratic system and to the ethic of a non-political military subservient to civilian control. Only the Sudan, because of its status as a state sponsor of terrorism, and Rwanda, because of substantial human rights violations, are prohibited currently from participation. In FY 1995 a total of $1.2 million afforded East African militaries an opportunity for short-term military education in the US. Countries participating included Burundi, Djibouti, Eritrea, Ethiopia, Kenya, Tanzania, and Uganda. A third, very small, Military-to-Military Program sends US officers to lecture and advise on how to reduce force structures, enhance military professionalism, and promote democratization and human rights adherence. These modestly funded programs help secure access for the US in those benefiting nations.

Parenthetically, both the British and French have maintained a military presence in the East African area, and although modest, is larger than that of the US. Since Kenyan independence in 1963, the British have had a defense pact with Nairobi and almost continuously kept a battalion of troops in country. However, because of President Moi's political intransigence, the British have allowed the agreement to lapse.[75] The French maintain approximately 3,000 troops in Djibouti, provide air, sea, and ground defense for the government, train the national military and police force, and underwrite substantially the government's national budget. Réunion, an overseas department of France, 4,000 kilometers from Djibouti, has been a French possession for over 300 years. It serves as headquarters for the French South Indian Ocean Armed Forces; approximately 3,000 French military are stationed on the island and are arguably essential to the local economy. Paris subsidizes 75 percent of the island's annual budget and trains the local forces in search and rescue, scientific missions, and civic action projects on the island.[76] The French sovereign presence is evident throughout the South Indian Ocean region.

Cultural Interests

President Clinton's address to the June 1994 White House Conference on Africa related that the two historic policy prisms through which the US viewed the continent were now shattered. Support for anti-communist and anti-apartheid governments is history. He urged the assembled participants to help develop an American constituency for Africa.[77] That has long been lacking.

Two areas have budding potential and have already begun to capture US public interest, sports and ecology. East African men and women have long captivated Americans with their Olympic track and field successes.[78] Near-legendary Ethiopian Abebe Bikila won the marathon in both 1960 and 1964. Mamo Walde, another Ethiopian, won the marathon gold medal in 1968. Other Ethiopians have gained fame as well. Miruts Yifter won the 5,000 and 10,000 meters in the 1980 games. Derartu Tulu was the first black African woman to win an Olympic gold medal in the women's 10,000 meter race at the 1992 games. In 1995 she won the world cross-country championship. Haile Gebr-selassie is the men's record holder in both 5,000 and 10,000 meters. Both Tulu and Gebr-selassie competed in the 1996 Atlanta games. Many East African sports stars now train in US sports facilities and, often, have US managers.

The second cultural area is the intergenerational US fascination with ecology and related tourism. The partial successes of East African governments in curbing the killing of the African elephant and rhinoceros for their tusks and horns have been well publicized. For the well-off US tourist, Kenya has long been the leader of 'sun and safari tourism' with its accessible wild game parks. Through a combination of eroding domestic infrastructure and overexploitation by visitors, Kenyan tourism has begun to shift toward Tanzania. From a US ecological tourism perspective, Tanzania is poised for rapid expansion of its game park visitors. Tanzanian President Benjamin Mkapa has made it clear that nature comes first and generating hard currency from tourism second. A major stimulant to tourism development has been the Aga Khan's $32.5 million investment in three resorts in Tanzania, all of which opened in June 1986.[79] These environmentally friendly resorts offer all the expected amenities. Each location proceeded through four environmental impact assessments resulting in the use of solar panels to produce hot water and a rooftop rainwater collection system to avoid dependence on a freshwater spring used by Masai herders. The government wants no more than 500,000 visitors each year in this fastest growing economic sector. In 1995, a record 293,834 foreign visitors spent $258 million in Tanzania's parks.

The many exchange programs that link Africa generally, and East Africa specifically, with US schools, museums, businesses, and American institutions, are a silent comment on this transnational connection. Afro-American tourism to the continent of their ancestry has often yielded follow-on civic, church, and youth exchanges. Celebrity presence of Americans resident in the region provide high profile advertising of the business and culture of East Africa. One example is the son of baseball star Jackie Robinson, who owns a 100-acre coffee farm in Tanzania.[80] On a small scale, one person at a time, there can be no better caring platform than the positive experience of US visitors to a region.

Institutional lobby groups such as the African-American Institute, TransAfrica, and the Congressional Black Caucus constitute the natural allies with the President and Congress to sustain presence and access to the region. While the American black community may provide the core of a Pro-Africa Coalition, it alone will not be sufficient nor uniformly motivated to do so. A broad and sustainable constituency can only develop if there are many common interests, reinforced over time. Private groups now supplement and could broaden further the necessary public opinion base within the US to complement the limited governmental resources likely to be devoted to the area. The Association of Returned Peace Corps Volunteers, Fulbright students from Africa, the US Chamber of Commerce, the American Newspaper Association, professional sports teams, the movie industry, and recording artists to name but a few, could help grow a basic American constituency to revitalize US interests here and elsewhere in the continent. As outlined above, East Africa's exposure to the viewing and reading US public has been rather remarkable and on balance there is a receptivity to know more.

Conclusion

For both external and domestic reasons, US interests and policies have been limited in the East African region and that posture is not likely to change soon. Along with unavoidable crisis-response measures sure to come, there exist some growth potential for certain states in the region and, at the same time, long-term benefit for US interests. Nevertheless, there is apprehension in the US Government and in the international community that without US leadership, success in democratization, sustainable economic growth, conflict mitigation, and African participation in resolutions of major transnational problems (drug trafficking, terrorism, AIDS, and environmental degradation) cannot be achieved. As the only remaining superpower, Washington may remain only reluctantly engaged to move these processes along. Multilateral responses will be the order of the day, whether prompted by political, economic, or military crises. Sustained official and private programs designed to deter and alleviate human or natural disasters meaningfully could allow time for democratization, shared political participation, and sustained economic development to become institutionalized in East Africa. Slippage from these goals can only result in even less US official and private attention to the region.

Notes

1. 'Hippos Sinking in a Quagmire of Problems', *National Geographic* (February 1996) 189(2), 139.
2. Blaine Harden (1990) *Africa; Dispatches from a Fragile Continent* (Boston: Houghton Mifflin), 130–2.
3. Eleanora West and Robert L. West, 'Conflicting Economic Interests of Africa and the United States', in Frederick S. Arkhurst (ed.) (1975) *US Policy Toward Africa* (New York: Praeger), 153.
4. A concise summary of missionary activity in the region can be found in Stanley Diamond and Fred G. Burke (eds) (1966) *The Transformation of East Africa* (New York: Basic Books), 513–33.
5. John A. Siewert and John A. Kenyon (eds) (1993) *Mission Handbook 1993–95*. 15th ed (Monrovia, CA: MARC, 15th edn), 40. Interestingly, in East Africa currently, the largest US missionary presence in total numbers, Protestant (1184) and Catholic (233) resides in Kenya. Other still sizable contingents can be found in Tanzania (400 combined total), Ethiopia (150), and Uganda (150). Major denominational presences include Southern Baptist, Seventh-Day Adventist, Church of Christ, Lutheran, Mennonite, the interdenominational African Inland Mission and SIM, and an array of Pentecostal groups.
6. Rupert Emerson (1967) *Africa and United States Policy* (Englewood Cliffs, NJ: Prentice-Hall), 18.
7. Cordell Hull (1948) *The Memoirs of Cordell Hull*, 2 (New York: Macmillan), 1235.
8. US Department of State (1971) *American Foreign Policy: Basic Documents, 1950–1955*, 2 (New York: Arno), 2173–4.
9. Chester Bowles (1956) *Africa's Challenge to the United States* (Berkeley, CA: University of California Press), 96–7.
10. An excellent survey of pre-independence, US governmental initiatives toward the conduct of African affairs is Vernon McKay, 'The African Operations of United States Government Agencies', in Walter Goldschmidt (ed.) (1963) *The United States and Africa* (New York: Praeger, rev. edn), 273ff.
11. Emerson (1967), Ch. 4, 'United States Interests and Activities in Africa', 29ff.
12. Andrew M. Kamarck, 'The African Economy and International Trade', in Goldschmidt (ed.) (1963), 157.
13. Emerson (1967), 33–4.
14. Ibid., 43, 46.
15. The most thorough treatment of the role of African militaries in this period can be found in W. F. Gutteridge (1969) *The Military in African Politics* (London: Methuen). Of particular note are his analyses of the impact of colonial heritage, ethnic favoritism, and the virtual absence, except in Ethiopia, of US military training support to East African states.
16. Kenneth W. Grundy (1968) *Conflicting Images of the Military in Africa* (Nairobi, Kenya: East African Publishing House), 12.
17. Examples include 'We Fight Rhodesian Independence but Bless Black African Dictatorships', *Congressional Record* (29 June 1966), A3493–4; and in the same source speeches by Congressman Ashbrook (OH) (8 March 1966), 5117–20; and Senator Eastland (MS) (25 August 1966), 19677–8, as cited in Grundy (1968), 38.

18. Emerson (1967), 102–3.
19. Herbert J. Spiro, 'US Policy: An Official View', in Arkhurst (ed.) (1975), 66.
20. Peter Woodward, 'Sudan: War Without End', in Oliver Furley (ed.) (1995) *Conflict in Africa* (London: I. B. Tauris), 98.
21. US law required the termination of diplomatic relations upon the occasion of a successful *coup*. In one of these interludes between *coups* I hosted the Sudan's Assistant Cabinet Minister for Youth and Sport, while he participated in a seven-week educational program in Columbus, Ohio in 1971. At that time, the minister was optimistic that US–Sudan relations would likely improve under Nimeri. That proved not to be the case.
22. Pierre Etienne Dostert (1994) *Africa 1994* (Harpers Ferry, WV: Stryker-Post, 29th edn), 202. Despite overwhelming evidence to the contrary, the Sudanese chargé d'affaires in Addis Ababa assured me in February 1995 that the civil war was well on its way to resolution, that the Sharia had not been forcibly applied to the Christian and animist populations in the south, and that ample food supplies were readily available in the region (Diplomatic Reception, US Ambassador's Residence, Addis Ababa, Ethiopia).
23. Office of the Coordinator for Counter-terrorism, US Department of State (April 1994) *Patterns of Global Terrorism 1993*, Department of State Publication 10136 (Washington, DC: US Government Printing Office), 25.
24. I encountered these concerns especially among Eritrean senior military officers and to a lesser extent by US embassy personnel in Asmara and Addis Ababa, February 1995.
25. I am indebted to my colleague, Joshua Rubongoya of Roanoke College, for bringing the Ugandan examples to my attention.
26. Eileen Alt Powell (2 February 1996) 'US diplomats fleeing Sudan', *Montgomery Advertiser*, np.
27. Associated Press wire story (Khartoum), (23 March 1996) 'Military leader wins election in Sudan', *Montgomery Advertiser*, 5A.
28. In the summer of 1970, immediately before Obote's downfall, I observed a society mired in statist bureaucracy and state-run enterprises, laden with tribal preferences, and severely detracting from an obvious potential for entrepreneurship, especially in the capital of Kampala. The chief exports, coffee and cotton piece goods, had to be sold to government-owned cooperatives at a fixed price, with frequent spoilage of the cash crop, little incentive to produce quality textile goods, and levels of government bureaucracy that stymied growth of an incipient economy. Observations as a participant in the Regional Council for International Education, Makerere University Faculty Symposia, Kampala, Uganda (June–August 1970).
29. The excellent first-person narrative of this episode appears in Robert V. Keeley, 'Crisis Avoidance: Shutting Down Embassy Kampala, 1973', in Joseph G. Sullivan (ed.) (1995) *Embassies Under Siege; Personal Accounts by Diplomats on the Front Line* (Washington: Brassey's), 13ff.
30. Ibid., 30.
31. Michael T. Kaufman (6 July 1976) 'An African Reaction: Public Anger, Private Approval', *The New York Times*, 3; (19 July 1976) 'The Fallout from Entebbe', *Newsweek*, 88, 43.

32. Dostert (1994), 185.
33. For one hopeful example, the World Bank reports a reduction in the number of state government ministries from 38 to 21. Civilian government employment has been cut in half, from 320,000 in 1989 to approximately 148,000 in 1995. At independence in 1962 Uganda exhibited one of the stronger economies on the continent, and in the ensuing years until Amin's *coup* the state reputedly had one of the most effective public administrations in East Africa. 'Uganda: Repairing a broken-down civil service', *Finance & Development* (September 1996) 33(3), 11.
34. See, for example, David S. Painter (summer 1995) 'RESEARCH NOTE – Explaining US Relations with the Third World', *Diplomatic History*, 19(3), 525.
35. Personal observation, driving tour of Asmara, Eritrea (February 1995).
36. Stockholm International Peace Research Institute (SIPRI) (1971) *The Arms Trade with the Third World* (Stockholm: Almqvist & Wiksell), 376. For a thorough and admittedly African sympathetic perspective of the Cold War impact on the continent, including much detail on East Africa, see Fred Marte (1994) *Political Cycles in International Relations; The Cold War and Africa 1945–1990* (Amsterdam: VU University Press).
37. For a concise discussion of these geostrategic myths, see Ambassador Daniel H. Simpson (15 August 1994) *US Africa Policy: Some Possible Course Adjustment* (Carlisle Barracks, PA: US Army War College Strategic Studies Institute), 4ff.
38. See Michael Clough (1992) 'Geopolitics and Africa', *Free at Last? US Policy Toward Africa and the End of the Cold War* (New York: Council on Foreign Relations Press), 10–13. Clough points out, by the mid-1980s a policy consensus emerged that the Soviet Union was pulling back from Africa and he even dates the end of the Cold War in Africa as 22 December 1988, with the signing of the tripartite agreement on Namibia, which included the withdrawal of all Cuban troops from Angola.
39. For elaboration of this major Soviet investment and the peculiar toing and froing of US–Soviet relations on the Horn, see I. William Zartman, 'Superpower Cooperation in North Africa and the Horn', in Roger E. Kanet and Edward A. Kolodziej (eds) (1991) *The Cold War as Cooperation* (Baltimore: The Johns Hopkins University Press), 156ff.
40. For a comprehensive appraisal of the impact of superpower militarization on this regime, see Balfour Agyeman-Duah (1994) *The United States and Ethiopia; Military Assistance and the Quest for Security, 1953–1993* (Washington, DC: University Press of America). While the thrust examines US militarization of the state through a long-term military assistance program, the post-1974 Soviet military aid program had the same effect.
41. 'US Relief Effort in Somalia'. Statement before the Subcommittee on Africa, House Foreign Affairs Committee, Washington, DC, 6 September 1992, *in US Department of State Dispatch* (28 September 1992), 738.
42. Associated Press wire story (Nairobi) (10 March 1996) 'Divisions in Somalia still deter needed aid', Sunday *Montgomery Advertiser*, 14A. In August, 1996 with more than a tinge of irony, Aidid's 31-year old former US Marine son returned to succeed his deceased father. He vowed to 'continue the policies' of the ex-president. Haroum M. Hassan (5 August 1996) 'Aidid's ex-Marine son chosen Somalia leader', *USA Today*, 1 IA.

43. Susan E. Strednansky, Maj, USAF (February 1996) 'Balancing the Trinity; The Fine Art of Conflict Termination', thesis, School of Advanced Airpower Studies (Maxwell AFB, AL: Air University Press), 26.

44. For a more positive assessment of the Somali operation, see Chester A. Crocker (May–June 1995) 'The Lessons of Somalia; Not Everything Went Wrong', *Foreign Affairs*, 74(3), 2ff. Formerly Assistant Secretary of State for African Affairs, his fundamental criterion is that 'we left it better off than we found it', still acknowledging that Somalia remains neither a peaceful nor hopeful place. An excellent article which appraises the implications for future UN peacekeeping drawn from Somalia is Paul F. Diehl (1996) 'With the Best of Intentions: Lessons from UNOSOM I and II', *Studies in Conflict & Terrorism*, 19(2), 153–77.

45. Quoted in Bruce B. Auster (13 March 1995) 'A Farewell to Distant Shores', *US News & World Report*, 6–7.

46. The Honorable Brian Atwood (26 July 1994) Administrator, Agency for International Development, 'Prepared Statement,' in *Crisis in Central Africa*, Hearing Before the Subcommittee on African Affairs, Committee on Foreign Relations, US Senate, 103d Congress, 2nd session, 8ff.

47. Associated Press Reports (12 October 1996) 'Burundi may need help, US says', *Montgomery Advertiser*, 6A.

48. The White House, *A National Security Strategy of Engagement and Enlargement* (February 1996) (Washington, DC: US Government Printing Office), 44. Indicative of 'déjà vu all over again', a virtually identical statement appeared in 'US Foreign Policy for the 1970s: The Emerging Structure of Peace', Report to the Congress by Richard Nixon, President of the United States (9 February 1972) (Washington, DC: US Government Printing Office), 104. The report maintained that 'If Africa is to move ahead . . . it must be largely on the basis of its own efforts.'

49. Thomas M. Callaghy, 'Africa and the World Economy: Caught Between a Rock and a Hard Place', in John W. Harbeson and Donald Rothchild (eds) (1991) *Africa in World Politics* (Boulder, CO: Westview), 42. Now the rule, rather than the exception, the British, French, German, Dutch, and Scandinavian governments have all imposed political conditionalities on foreign aid with the US first among donor nations in so doing. Earl Conteh Morgan (fall 1994) 'The Military and Human Rights in a Post-Cold War Africa', *Armed Forces and Society*, 21(1), 76.

50. Patrick Clawson (ed.) (1995) *Strategic Assessment 1995; US Security Challenges in Transition* (Washington, DC: US Government Printing Office), 97–106.

51. Anecdotally, in a 1995 trip into Eritrea, both in the capital of Asmara and the chief port of Massawa, signs of a positive popular outlook, commitment to sacrifice, and rebuilding infrastructure from the war were widely evident. That view was confirmed in briefings by Ambassador Houdek and members of his country team. In contrast with a much earlier country visit in 1973, however, clearly the war has taken a major toll on the physical infrastructure, both in terms of ordnance damage (exploded and live, uncleared landmines and the destruction of all banks in Massawa for example) and no monies for civic maintenance.

52. Kim R. Holmes (ed.) (June 1994) *A Safe and Prosperous America; A US Foreign and Defense Policy Blueprint* (Washington, DC: The Heritage Foundation, 2nd edn), 19, 36–8, 48.
53. Samuel P. Huntington (summer 1993) 'The Clash of Civilizations?', *Foreign Affairs*, 72(3), 22. While his thesis has much to say for it, that these 'fault lines' are the battle zones of the future, universal application is quite another conjecture.
54. This point is well made in Asteris C. Huliaras (December 1995) 'Islam: promise or peril?', *The World Today*, 51(12), 242.
55. Jennifer Seymour Whitaker (1996) 'Africa: Should the US Care?', *Great Decisions 1996* (New York: Foreign Policy Association), 65.
56. Telephone interview, Deputy Director for East African Affairs, Department of State (8 March 1996). The immunity grant was reported in 'Seychelles' Offer Castigated', *The Wall Street Journal* (2 February 1996), A6.
57. Whitaker (1996), 70–1.
58. *Amnesty International Report 1994* (New York: Amnesty International, USA), 182–4.
59. Smith Hempstone (winter 1995–6) 'Kenya: A Tarnished Jewel', *The National Interest*, no. 42, 55.
60. US Department of Commerce, Economics and Statistics Administration, Bureau of the Census (September 1995) 'Table no. 1330. US Government Foreign Grants and Credits, by Type and Country: 1946–1992', *Statistical Abstract of the United States 1995*. (Washington: US Government Printing Office, 115th edn), 810.
61. US Department of Commerce, Economics and Statistics Administration, Bureau of Economic Analysis (August 1995) 'Table 17 – US Direct Investment Abroad: Country Detail for Selected Items', *Survey of Current Business*, 75(8), 97, 114.
62. Steven Metz, 'Africa', in Earl H. Tilford, Jr. (ed.) (1 February 1996) *World View: The 1996 Strategic Assessment From the Strategic Studies Institute* (Carlisle Barracks, PA: US Army War College Strategic Studies Institute), 38; Marguerite Michaels, 'Retreat From Africa', *Foreign Affairs*, 72(1), *America and the World 1992–1993*, 6.
63. Lee Michael Katz (8 October 1996) 'Christopher makes a statement with trip to sub-Saharan Africa', *USA Today*, 12A.
64. Richard W. Hull (May 1991) 'The Challenge to the United States in Africa', *Current History*, 90(556), 193.
65. Craig Johnstone (16 October 1995) Director, Resources, Plans, and Policy, Department of State, 'Foreign Policy on the Cheap: You Get What You Pay For', Address before the Seattle World Affairs Council, Seattle, Washington, 6 June 1995, *US Department of State Dispatch* 6(42), 743.
66. 'A Thirty-Year Perspective: Past and Future' (1989) *Sub-Saharan Africa, From Crisis to Sustainable Growth* (Washington, DC: The World Bank), fig. 1.2, 18.
67. 'Rich Nations Endorse Plan to Relieve Debts', Sunday *Montgomery Advertiser* (29 September 1996), 14A.
68. For an especially thorough discussion of the contributions and constraints of NGOs and associational practice in Africa, see Naomi Chazan (spring 1992) 'Africa's Democratic Challenge', *World Policy Journal*, 9(2), 279ff. In Addis Ababa there were 60 registered NGOs working in country, and

another 30 to 40 not registered in early 1995. Interview, US embassy official, February 1995.

69. For an early appraisal of tourism's trade-offs for Africa, including examples from East Africa, see Paul E. Jursa and James E. Winkates (spring 1974) 'Tourism as a Mode of Development', *A Quarterly Journal of Opinion*, 4(1), 45ff.

70. 'DOD Strategy for sub-Saharan Africa' (1996) Issue 6, *Defense 95* (Alexandria, VA: American Forces Information Service), 38. An excellent survey of constructive, peaceful, US military contributions to the continent, much of which would readily apply to East Africa, can be found in Commander Robert W. Higgs South African Navy, 'The United States and Sub-Saharan Africa', *Naval War College Review*, 49(1) (winter 1996), 90–104.

71. The 20-year-idle rail line connecting Massawa to Asmara is being rebuilt entirely by manual labor because the fledgling government cannot afford to hire modern contract railroad engineers. The rail bed rises 8,000 feet over 73 miles through steep gorges and mountain passes. Resurrecting the rail line has become a metaphor for Eritrean nation building. James C. McKinley, Jr (19 May 1996) 'Working on the Railroad, and on Eritrea's Revival', *The New York Times*, 3. Between Assab and Massawa lies approximately 300 miles of very rough road, with luck a 12-hour drive; it takes another 6 hours to reach the capital. Conversation with an Eritrean driver, February 1995.

72. 'Tracking Station Shuts its Doors' (September 1996) *Guardian* (Official Magazine of the Air Force Space Command), 4(3), 18. The Indian Ocean Tracking Station was inactivated after 33 years of satellite support for NASA and Department of Defense (DOD) missions.

73. The information on security-assistance programs presented appears in some detail in Maj. Joanne Bernstein, USA (fall 1994) 'Military Assistance in sub-Saharan Africa', *The DISAM Journal of International Security Assistance Management*, 17(1), 90ff.; and (spring 1995) 'The Security Assistance Budget Request for FY 1996', *The DISAM Journal*, 17(3), 54ff. My own institution, the Air War College, educates 40 international officers, including a small number from Africa, annually in a ten-month graduate-level program.

74. 'Table No. 1332. U.S. Foreign Military Aid, by Region and Selected Countries: 1985–1993,' *Statistical Abstract of the United States 1995*, 813.

75. Hempstone (winter 1995–6), 57.

76. Michel Klen (January, 1996) 'Quel avenir pour la Reunion?', *Revue Defense Nationale*, 149ff.

77. Excerpts of the major conference addresses can be found in 'The White House Conference on Africa' (July 1994) CSIS Notes, no. 162 (Washington, DC: Center for Strategic and International Studies). The full text of the President's speech appears in 'Remarks to the White House Conference on Africa, 27 June 1994' (1995) Public Papers of the President, William J. Clinton 1994. Book I, 1 January to 31 July 1994 (Washington, DC: US Government Printing Office), 1150.

78. Jere Longman (3 July 1996) 'Running With a Continent's Hopes; Tulu Seeking a Repeat Gold "for Women in Ethiopia and Africa"', *The New York Times*, BI 1.

79. Hrvoje Hranjski (25 August 1996) 'Tanzania's ecology comes before big money of tourism', Sunday *Montgomery Advertiser*, 2H.

80.   Sharon Raboin (26 September 1996) 'Robinson's widow offers 'Intimate Portrait' of Jackie', *USA Today*, 3C. Rachel Robinson, Jackie's widow, finished a ten-city tour publicizing her new book, *Jackie Robinson: An Intimate Portrait*, in which she recounted her son's Tanzanian farm ownership. She founded and chairs the Jackie Robinson Foundation, whose purpose is to provide financial support and leadership development to minority students.

5
Southern Africa: US Relations in Transition

Karl P. Magyar

Introduction

Histories of Africa often allude to Pliny's ancient observation, 'Out of Africa, always something new.' For Pliny, the African world consisted of the continent's northern region, but in our own day his observation is especially appropriate for southern Africa which mistakenly enjoys a relatively more tranquil and orderly reputation among American observers than does the rest of the continent. While West, Central, and East Africa are popularly portrayed by academic and media analysts and, even by some US government agencies, as chaotic and marginalized in the post-Cold War era, certain countries in southern Africa are viewed as politically promising and budding democracies, and in economic terms some have been labeled 'emerging markets'.[1] However, among some analysts who follow southern African affairs, contrary perspectives are offered. Michael Clough reported, '*southern Africa* – has been a battleground for much of the post-independence period. While colonialism crumbled elsewhere in the 1960s and 1970s, in Angola, Mozambique, Namibia, Rhodesia (Zimbabwe), and South Africa, white rulers clung tenaciously to power. The result was long and bitter wars that engulfed the entire region and drew in the superpowers'.[2] And Christopher Coker has noted, 'Perhaps the most tragic region of all is southern Africa, which between 1980–88 saw 1.9 million deaths in war, 8% of the population of Mozambique and 7.5 % of Angola. . . . 14 million refugees were produced, half the population of Mozambique and Angola combined.'[3]

The gap between external popular perceptions regarding southern Africa, and the realities, may perhaps be best explained by the presence of substantial numbers of whites traditionally residing in this region,[4] especially in South Africa, whose domineering power over the region

139

implied greater political and social stability and advanced economic achievements to many external observers. While the latter characteristic is generally valid, the presence of the whites, however, was also a major contributing factor in the region's volatile character. In contrast to West Africa, where historically there were indeed very few European settlers, the vast fertile lands, abundant natural resources, and the moderate climate of southern Africa attracted sizable European populations, many of whom had staked their claims to large chunks of land, and for whom the non-white populations were little more than cheap and readily available labor reserves.

Where West African countries attained independence for the most part peacefully and abruptly, whites in southern Africa tenaciously sought to retain their hard-won domination over the land and its native populations. And where the rest of sub-Saharan Africa came quickly under the control of indigenous black populations during the rush to independence, the minority white rulers of the lands of southern Africa resisted several major militant attempts by majority black populations to come to power. With some historical irony, the whites won most of the armed battles, but lost all the wars. In the end, blacks came to power in every country of this region as well. This is true also of South Africa where black militant opposition only modestly challenged the government's ability to maintain control. Yet despite the ruling white Afrikaners' powerful and well-tested military machine, rated as the best in the globe's entire southern hemisphere, that government yielded voluntarily its domination in response to a variety of deteriorating political, demographic and economic factors.[5]

Many interesting developments contributed to the rapidly changing events in this volatile region. The Portuguese had demonstrated an audacious attempt by a waning and poor European colonial power to retain control over Angola and Mozambique whose combined geographic area was 22 times the size of Portugal. In Rhodesia, whites attempted to retain control over a black population 35 times the size of their own, and of those 200,000 whites, one-half were recent immigrants or only temporary residents with little history of loyalty to the colony.

The considerable riches of natural resources such as oil in Angola's geographically detached Cabinda province, and the mineral belt stretching from Zaire's southern Shaba Province, through Zambia, Zimbabwe, Botswana, and into South Africa enriched Europeans such as Cecil Rhodes and numerous London financiers, as well as many white settlers. But the dislocated native populations, whose physical labor was essential for extracting fortunes from those mines, became the source of modern political unrest and subsequent activist challenges to colonial or minority rule. Nowhere was this illustrated better than in the black townships of South Africa such as Soweto (adjacent to Johannesburg), which were originally erected to contain the initially docile labor forces, but became the breeding grounds of escalating political opposition to white domination. South-West Africa (which became Namibia) had the misfortune of falling into South Africa's hands as the result of the defeat of its colonizer, Germany, in World War I. That vast territory finally attained independence in 1990 as a by-product of the Soviet Union's internal demise and the subsequential abandonment of its substantial, but largely incomprehensible, political and military commitment in Angola. Domestic Marxists were successfully maneuvering themselves into power in Angola, Mozambique, Zimbabwe, Namibia, and South Africa, yet after doing so, in every case the orthodox Marxist operational structures were soon abandoned and certainly divorced from an internationalist, conspiratorial context. Perhaps the supreme paradox in this perplexing region was that of South Africa's surprising reversal of apartheid-based, authoritarian control by whites over non-whites: at a time the Soviets and Cubans retreated from the African continent, leaving white control over South Africa unchallenged and in a position of strength, the whites abandoned their apartheid legacy and allowed blacks to come to power in the first democratic, all-race, 1994 election.

In short, southern Africa's volatile nature derives from the dynamic interaction of geography, diversified races and cultures, a substantial white presence, huge quantities of minerals, Marxists, nationalists, extracontinental intervenors and financial interests – and above all, the domination of the entire region by white-run South Africa. Much blood was shed historically, and (except during the Anglo-Boer War, 1899–1902, and some skirmishes during World War I,) almost all of it the blood of blacks. Great fortunes were made – mostly by whites. Yet there exists today wide optimism by external observers about the future promises of developmental prospects of all of South Africa's population segments, which is to ensue from the stabilization of the entire region. Again, this stands in contrast to the rest of sub-Saharan Africa whose marginalized condition is viewed by many as likely a lasting feature.

For the United States, some major adjustments in its relations with this region have been underway since the end of the Cold War. Traditionally, one-third of America's trade with the continent of Africa was with South Africa. However, all mutual economic relations with that country became increasingly controversial during the waning days of apartheid in the late 1980s, when stepped-up external economic sanctions demonstrated their effectiveness. America's other interests in this region concerned the extensive activities of Cuba and the Soviet Union, especially in Angola, and aid supplied by various other extracontinental Marxist or other ideologically driven states to insurgencies, or to established, but weak, governments. The United States quickly and unambiguously accommodated Nelson Mandela as President of South Africa; abandoned support of Jonas Savimbi's UNITA insurgent movement in Angola in favor of the government forces effectively in power; openly welcomed Sam Nujoma's election as President in Namibia; and actively supported the government in power in Mozambique – from which the United States had kept a calculated distance due to its previous Marxist orientation and ties to external Marxist states.

Geographic and Demographic Features

Most accounts of southern Africa include ten countries, stretching from Angola, Zambia, Malawi, to Mozambique, and the six other countries to the south. Occasionally, Tanzania is added to this region, but that country will not be included in southern Africa in this review. In terms of area, southern Africa's size is almost equal to two-thirds that of the United States (including Alaska). In comparison to Europe, southern Africa's area is nearly four times as large but contains only one-quarter of Europe's population. Southern Africa has been defined by the modern political boundaries of the constituent states, but if geographic features and economic activity were to be the main criteria, certainly Zaire's southern Shaba province (known at times as Katanga) would be included as that province is distantly removed from Zaire's Atlantic Ocean orientation and location of its capital city. Most of Shaba's economic ties, centering on mining industries, are integrally related to southern Africa's rich geological zone, and economic and transportation infrastructure. Generally, more than half of Shaba's trade flows through South Africa's ports, after traversing Zambia and Zimbabwe. During its long period of destabilization in neighboring affairs, South Africa had supported if not engineered, the destruction of railway facilities in Angola and Mozambique, thus forcing the six landlocked countries in this region to ship through South Africa. This had proved to be an extremely effective political strategy as it had exacted compliant behavior from those countries who might otherwise have accommodated increased hostile activities by the African National Congress (ANC), the major opponent of the white regime, from bases within their borders. The United States expressed great sympathy to the victims of South Africa's destabilizing measures, but this unfortunate episode had to be weighed against the substantially sized Soviet and Cuban commitments to the region – especially to Angola.

A main feature of the southern African region, and one which may play a major role in the future, is geographic homogeneity and a terrain which poses no physical obstacle for the potential progressive integration of economic features and infrastructures of all the region's states. A geographic barrier is not encountered until reaching the vast tropical forested zone of central Zaire (renamed 'Congo' once again). As neighboring countries are unable to build or to maintain the vital transportation infrastructures necessary for the export of minerals and a great volume of imports, South Africa's modern and well-maintained infrastructural facilities cannot be obviated. This contributes immeasurably to that country's power and controlling authority. This was amply illustrated when Zimbabwe supported at great expense its own military forces charged with physically guarding the railway from Harare to the port city of Beira in Mozambique in order to secure its trading links with the rest of the world. South Africa's ports and railway links were well maintained and available to Harare, but Zimbabwe reasoned that the political instability within South Africa could eventually disrupt its transport through that country and hence the expensive alternate route through Mozambique had to be maintained. Mozambique was at the time in the midst of its own civil war and therefore unable to guarantee the security of its own facilities. Such sobering realities were demonstrated in 1986, when South Africa assisted (if not engineered) the removal of Lesotho's government led by Chief Jonathan, in a *coup d'état*.[6] South Africa assisted by simply halting all shipments destined to that landlocked country at Durban's port.

South Africa's traditionally dominant position in the region is a well-established feature. With about 40 percent of the region's population, South Africa produces well over 80 percent of the region's combined GNP. That country generates some three-quarters of all regional exports, and maintains a very favorable trade balance with Africa, mostly with southern Africa. South Africa produces in excess of 75 percent of the region's coal, iron, several strategic minerals, electricity, maize, sheep, meat, and motor vehicles. South Africa also possesses 57 percent of the region's operating railway lines, 63 percent of tarred roads, and 84 percent of telephones.[7] By contrast, South Africa's nearest competitor in the economic arena, Angola, generates only about 7 percent of the region's combined GNP. In terms of trade, South Africa generally exports three times more to the countries of southern Africa than she imports from them. These represented data for South Africa under white rule despite the impact of greatly restrictive international punitive economic sanctions.

Under peaceful majority black rule, with foreign capital flowing modestly back into the country, South Africa's regional hegemonic position may yet be enhanced. In fact, prior to majority rule in South Africa, that country held an overwhelming dominance in the regional military and economic arenas, and this is expected to increase in the future. Put simply, whichever regional institution South Africa joins, South Africa will automatically occupy the controlling position. The other countries in the region may play marginal, and now, increasingly diminishing political roles and to a large extent, for the foreseeable future, the economy of southern Africa will in effect mean South Africa. South Africa's domineering geo-economic feature had posed a perplexing dilemma to the other members of southern Africa during white rule, but this overwhelming character even under black rule will no doubt prevail in any new regional structures. The United States has on numerous occasions expressed the expectation of furthering black-led South Africa's responsible leadership role in the region. This, of course, has sensitive implications for US relations with South Africa's neighboring countries who are apprehensive of South Africa's potential hegemonic aspirations. Botswana and Zimbabwe are early examples. In no other region of Africa is there such a parallel. In West Africa, Nigeria had at one time been identified by external powers, including the United States, as a potential regional leader. However, stemming largely from the ongoing military domination of that government and its failure to reinstate democratic institutions in Nigeria, this has prevented the United States from publicly acknowledging that country's regional role as stabilizer. In southern Africa, there is simply no alternative to South Africa assuming this function. This has been openly acknowledged by the United States who lauds the virtues of evolving regional democratic institutions, regional stability, and regional peacekeeping mechanisms.

Demographically, South Africa, with over 40 percent of southern Africa's population, dominates the region. Noteworthy is a fact that will perpetuate South Africa's prime role, that being the low population figures for all other southern African countries. It must be understood that the whites who controlled power in South Africa constituted less than 5 percent of the combined regional population which, in light of South Africa's previous domination of the region, meant that the small white minority virtually held the rest of the region hostage.[8] South

Africa's nearest population numbers competitor is Mozambique with some 15.7 million inhabitants. Angola, Malawi, Zambia, and Zimbabwe contain between 8 and 10 million each, while Botswana, Lesotho, Namibia, and Swaziland each count less than 2 million inhabitants. Six of the region's countries have a common border with South Africa that again highlights that country's physical controlling position. This latter feature played an important role when the white government exercised unchallenged destabilizing measures in hot, cross-border pursuit of the ANC. Four countries are microstates in terms of their population numbers, while all the rest, except for Mozambique, qualify as small states. This characteristic will further impede their growth prospects due to their inherently small and generally poor markets, which will not attract much foreign private developmental resources beyond satisfying only modest marketing opportunities. South Africa will be in a favorable competitive position in the region vis-à-vis the United States due to historical economic ties, the region's most advanced level of technology, and an ideal geostrategic position.

South Africa's comparatively large population base and overwhelming economic presence will make that country the only focus of global economic attention, relegating the fortunes of the other states to servile or marginal positions. Botswana may expect a somewhat better fate due to its high rate of economic growth for over a decade, and this has drawn some American interest. But that country's narrow mineral base, which fuels that growth, is subject to international market vicissitudes and to South Africa's influence as South African expertise dominates the development, extraction, and transportation of Botswana's minerals – the base of most of its ample wealth. Only South Africa and Mozambique have populations larger than that of New York city, hence when projecting economic development and interaction between the members of this region, and their scarce potential for attracting future foreign investments, the inherent geo-economic limitations of most of the other members of the region must be appreciated. And as a final precaution, as is typical of the rest of Africa, the internal income distribution in each of the countries of this region is greatly skewed, with much of it concentrated in the hands of a small segment of their respective populations. Private United States developers will naturally prefer to establish their investments in South Africa due to its extant modern marketing and industrial infrastructure, including harbors, roads, and excellent transport facilities. As such, American entrepreneurs will be re-identified as South Africa's economic and political ally by the other members of the region. Arguments may surface that only political matters have changed in the ascendancy of the black majority to power, but in the vital economic realm, America's meaningful inter-action remains mostly with the wealthy white elite in South Africa, and possibly with the small ruling oligarchies in the rest of this African region.

Europe's Regional Penetration

The major historical features of human activities in southern Africa concern similar affairs that also characterize the rest of the continent. The indigenous population which traditionally ranged over a wide area were the Bushmen (San), and in the present Cape area of South Africa, the Khoi-Khoi (Hottentots). (Together, in scholarly literature, they are now referred to as the 'Khoi-San'.) The San have greatly diminished in numbers and exist today only in concentrated pockets in Botswana, Namibia, and Angola. The small numbers of Khoi-Khoi and their numerous descendants, who mixed with whites, are to be found in South Africa and Namibia. Two major migrations of Bantu arrived in southern Africa, one along the Atlantic coast which included, among others, the Ovimbundu of Angola and northern Namibia, as well as the Hereros of north-central Namibia.[9] The latter comprise the southern-most Bantu people who had migrated southwards. The other Bantu migrations arrived along the Eastern highlands of Africa and settled along the coast and interior of southern Africa. These Bantu represent primarily the Nguni and Sotho linguistic strains and include, notably, the Xhosa as the southernmost migrants along the eastern seaboard of South Africa, the Zulus and Sotho and their related tribesmen, in addi-tion to several smaller tribes, such as the Venda and Tsonga, judged to be of separate origins.

This variety of indigenous African peoples was reason enough for many conflicts over land in the southern Africa region, but then the Portuguese explored and settled along the coasts of Angola and Mozam-bique. Later European settlers were the Dutch, starting in the 1650s, who arrived in the Cape region of South Africa – the lands of the Khoi-Khoi. British imperial controls in southern Africa 'waxed and waned' in the words of Ieuan LL. Griffiths, reflecting 'changes in government at Westminster and swung with the fortunes of Tories and Liberals'.[10] The British took control over the Cape during the Napoleonic wars at the end of the eighteenth century and subsequently extended imperial rule all the way through South Africa into Rhodesia. British Protectorates were also established over what are today Botswana, Swaziland, and

Lesotho. Still another European population in southern Africa was the Germans who colonized Namibia. To this kaleidoscope of people, the British also introduced the Indian population into South Africa – where today they constitute a politically very visible, but vulnerable, 3 percent of that country's population. And, in turn, even this cultural community is also divided by ethnic differences.[11]

The implications of this tumultuous history of migrations is the continuous and clashing claims that confronts the external interests such as the United States with a lack of consensus in the attempt to formulate lasting diplomatic policies. Our vacillating history of policies with respect to extending support to Angola's insurgency leaders Holden Roberto and Jonas Savimbi testify to this dilemma. This is especially acute where distinct ethnic social sectors represent established political cleavages such as President Mugabe's Shonas in Zimbabwe, or President Nujoma's Ovimbundus in Namibia. The Soviets had gambled recklessly when they supported Joshua Nkomo's Matabele-based insurgent movement – which represented but one-fifth of Zimbabwe's native black population. For the United States, a similar Reagan-era policy dilemma concerned the overt acknowledgment of the infinitely ambitious Chief Gatsha Buthelezi, the leader of the white government-established Kwa-Zulu homeland, who was hounded by the Xhosa-dominated African National Congress from whence Nelson Mandela drew his power. By contrast, President Clinton soon changed policy towards Angola where the long-beleaguered government was finally recognized by the United States; and towards South Africa, where the all-party elections wrested power from the whites.

Virtually all groups of indigenous or immigrant peoples experienced organized violence as victim or aggressor at some point, hence much of the modern conflict episode is but a continuation of historical conflicts in this region. The major object of contention then was land, where today it is political power and authority over those who wish to perpetuate their domination or historical claims to the lands. There exists today externally a mistaken popular portrayal of armed conflicts in this region, which until the all-race election in South Africa in 1994, were assumed to be between blacks and whites. That was the case during the initial stages of Angola's, Mozambique's, and Namibia's liberation wars, and in Zimbabwe's and South Africa's majority black struggles against European white colonial immigrants who appropriated vast tracts of land. However, the longest wars and those in which the greatest number of casualties were inflicted, were armed conflicts by blacks against each other.[12] After attaining independence, Angola and Mozambique's internal factions kept fighting for another two decades, with Angola's civil war still not resolved decisively to this day. In Zimbabwe, post-majority rule saw clashes between government troops, now under the control of President Robert Mugabe, and elements of Joshua Nkomo's Zimbabwe African People's Union, continue for a year. In South Africa, all but a small portion of modern-day politically-induced casualties occurred from black-on-black violence, and most of these were from opposing factions among Zulus who probably yielded around 90 per-cent of all South African politically related casualties on opposing factions of their own ethnic group. Their battles continue to this day despite majority rule – though at a lesser level. Of course, much of the violence among blacks in southern Africa had its origin in the disruptions initially introduced by white intrusion, and subsequently in struggles against white domination. But this violence had been ignobly continued, not without great enthusiasm, by the internal struggles for power among the majority blacks who now control all the governments in the region. For the United States during the Cold War this implied that some tough choices had to be made between support of the known white colonial or minority regimes, and unknown black challengers, with the latter purportedly seeking self-determination – a hallowed democratic value. However, the contest was played out in a peripheral region of the Cold War, hence ambiguity and inconsistency were tolerated and even institutionalized. The official circles in the United States had easily argued that no direct confrontation with the Warsaw Pact states should be risked in this distant and relatively unimportant region.

Southern Africa: The Cold War Context

Having highlighted southern Africa's broad geographic, demographic, and historical dimensions, we may now review America's interests in this region during the period of the Cold War. Several factors may be readily identified as comprising the major variables that combined to shape America's policies in pursuit of those influencing interests in southern Africa. In synoptic (but not necessarily hierarchic) listing, those influencing factors were the region being part of a volatile continent undergoing a fundamental political transition; the vast colonial holdings in southern Africa of our NATO ally Portugal; white-controlled South Africa's highly controversial internal ruling policy of apartheid; the region's trade value and source of 'strategic' minerals; the allegedly 'vital' international shipping lanes that must be kept open at any cost; South Africa's uncontested military, political, and economic domination of the region; Marxist-influenced and externally supported liberation wars; and substantial but nebulous Soviet and Cuban interests and meddlesome activities in the region that required close monitoring and neutralization.

These factors served to channel America's focus on southern Africa – much as they had attracted the activist intervention of the Soviet Union. A surrogate Cold War military battlefield had been introduced in Angola in whose post-independence civil war external intervention ceased only with the collapse of the Soviet Union, the result of that empire's own internal collapse. Deprived of Cold War sponsors, however, the battles in Angola continued but within a new political context that included negotiated settlements, elections, and United Nations peacekeeping activities.

Of Africa's five regions, southern Africa was last to experience fundamental political transformation, whether independence or majority rule. Malawi and Zambia had attained independence in 1964, but four years earlier, it was the Central African region's Belgian Congo's abrupt and violent transformation to independence in 1960 that had caught the United States by surprise. (This Congo became known as Zaire, and is not to be confused with neighboring Congo, Brazzaville.) The Soviets had taken an early interest in Zaire whose first Prime Minister, Patrice Lumumba, and one of his successors, Antoine Gizenga, had invited Soviet involvement against their Western-backed opponents.[13] In a flurry of political confusion, the United States backed Joseph Mobutu, Zaire's military head, who ousted both, the feuding civilian President and Prime Minister, and took over power a few years after attaining independence.[14] Except for a brief interlude, Mobutu ruled the country until mid-1997, having been propped up at great political, as well as financial expense by the United States.[15] This was done initially to bring order to this vast, secessionist-prone and anarchic land that had caught the attention of the Soviets, and because of Zaire's proximity and availability to service Jonas Savimbi's UNITA insurgent activities against the Soviet- and Cuban-backed Angolan government. Though a part of Central Africa, Zaire has frequently figured in United States policy in the affairs of southern and eastern Africa, despite Mobutu having been an embarrassing ally. Immanuel Wallerstein asserts persuasively that it was America's involvement in Zaire that 'was the launching pad' of the new American attitude towards southern Africa.[16]

Although 17 sub-Saharan countries attained independence in 1960, they were mostly countries in West and Central Africa, far removed from perceived vital and measurable interests in southern Africa. To

South Africa, the impact of the newly attained independent status of these states in West and Central Africa failed to be as profound as was the Congo's transformation. The 1960s, however, was also a particularly active era of the Cold War in Europe and the Far East, and while the United States could not divine what Soviet interests in Africa might be, America's Cold War vigilance policy dictated that Soviet political and diplomatic offensives in Africa needed careful monitoring – as well as assertive neutralization. The Soviets were politically and economically ill equipped to intervene massively in Africa, hence they extended only a probing, adventurous, and politically supportive agenda to Africans who had had no prior experience with that Cold War power.

Ironically, neither the United States nor the Soviet Union ever had colonies in Africa, hence each acted largely out of ignorance of the continent's complex developments. Each greatly overestimated the continent's potential importance to the other in the global strategic and economic sweepstakes; each saw Africa as a Cold War prize; and each ultimately invested substantial economic and political sums in Africa – with the Soviets ultimately realizing no appreciable return as the direct result of involvement. Both also were attracted to southern Africa after brief flirtations with other regions, which, however, soon revealed their innate political shortcomings. It was the Soviets who had usually set America's agenda in much of sub-Saharan Africa in that the United States mostly reacted to Soviet initiatives, and this was also very much the case in southern Africa. There were isolated pockets of interest in southern African affairs in Washington, but throughout the Cold War years, the White House had regarded the African continent as a low political priority except for keeping the Soviets from spreading their aggressive meddling in southern Africa.

Since the end of the Cold War, the United States systematically gained wide respect throughout most of sub-Saharan Africa. But in the absence of the Soviets, this affords no great advantage to the United States, as the consequences of such intensified relations would inevitably lead to bottomless requests for financial assistance in competition with other strategically more crucial global areas of concern. On receiving the new Ambassador from Comoros, in 1997, President Clinton meekly expressed the reason for the United States closing its embassy in Comoros as stemming from domestic budget cuts.[17] More likely, the reason concerns the lack of any identifiable interest to halt the progressive marginalization of much of Africa. And except for a modest amount of trade and somewhat intensified economic relations between the United States and South Africa, after the introduction of majority rule in that country, these low-level commercial characteristics also describe relations with the other states of southern Africa.

World War II had played a major role in the subsequent transformation of Africa's colonies into independent entities. Europe's colonial powers had weakened during the war, and the colonies were soon judged to be excessively burdensome. Only Portugal would persist in maintaining her vast but waning imperial extension in southern Africa, arguing that her African holdings were overseas provinces of metropolitan Portugal and not colonies.[18] In the late 1950s an externally influenced, orthodox guerrilla movement sprang up in Portuguese Guinea in West Africa, which fought for, and attained independence as Guinea-Bissau in 1974. National liberation movements in Angola and Mozambique began in 1960, and they attained independence from Portugal in 1975. Each of these colonies saw the major liberation movement influenced by Soviet or other Marxist forces and hence they caught the attention of the United States. Angola was a special focus for the United States as several administrations weighed Angola's Soviet factor against United States domestic Cold War anti-Communist considerations.[19] Marxist governments or Marxist-influenced guerrilla opposition movements covered a large geographic part of southern Africa, with only Rhodesia's and South Africa's white governments having waged meaningful opposition. America's reluctant alignment with the fiercely anti-Communist white South Africa regime implied a pragmatic policy to at least tolerate temporarily apartheid de facto, although it was officially and publicly castigated. In stark operational terms, the choice in southern Africa was between alignment with atavistic-tending whites, representing the past and present, and blacks, potentially in alliance with the trouble-some Soviet-Cuban axis, representing, perhaps, the future.

The collapse of the Portuguese colonies had validated the prescient warnings to South Africa in 1960 by British Prime Minister Harold Macmillan of the 'winds of change' sweeping the African continent. Those winds had toppled colony after colony throughout the continent, and white-run South Africa soon perceived a red and black noose tightening around her neck. South Africa's Afrikaners termed it the *Rooigevaar* – the 'red fear'.[20] There was some sympathy for this situation among certain American officials, and it realized its most advanced expression in the Reagan Doctrine, which overtly extended support to active Third World opponents of hostile Soviet-supported regimes. Angola was Africa's most notable demonstration of the focus of this doctrine.[21]

As rapidly changing events in southern Africa began to unfold, the United States experienced the true complexity of operating in sub-Saharan Africa, a sizable chunk of the world previously unknown to most Americans. The time period of these events in Africa coincided with the perplexities of the Vietnam War, which, along with strategic threats in East Asia and Europe, were infinitely more important contests of the Cold War. Except for arresting Soviet political advances and gains in Africa, and assuring access to southern Africa's strategic minerals, the United States simply had no vital interests in black Africa that would have justified a major commitment of human and military resources. For the United States, Africa primarily became the focus for humanitarian interests. What the Soviets had hoped to gain in Africa was beyond comprehension and awaits a definitive elucidation.[22] If East Asia and Eastern Europe represented America's major Cold War chess opponents, only a few civilian pawns were ever engaged in Africa in order to forestall unforeseen Soviet designs. In other words, some intelligence efforts, political support, and very modest development sums were invested without great consequence. Evidently the most sensitive location of such competitive attention was Angola, whose strategic location could possibly become the most likely regional base of any future concerted violent attempt to topple South Africa's white government, for intervention in Zaire, or, as has been alleged, that Angola may be the best location for monitoring or servicing potential armed hostilities in the southern Atlantic region. Angola is the third largest member of the region, and is often identified as having the second largest economy in southern Africa, but a very distant second. However, none of these arguments lack severe critics.

During President Nixon's tenure in office, Secretary of State Henry Kissinger directed that a study be made by various government agencies which would offer options for American policy in southern Africa that would avoid having to make a choice between whites and blacks. Liberated black Africa should not be offended by United States policy, but there were certain American early Cold War interests that could not simply be disregarded in white-run areas such as Angola, Rhodesia, and South Africa. The resultant classified report *National Security Study Memorandum 39* (*NSSM 39*), which soon saw open publication, in effect argued that support for Portugal and for South Africa should be extended as they were our natural allies against communist adventurism in Africa. Portugal's employment of military weapons originally supplied to that country in line with NATO's objectives in Europe, but which had been utilized in its African colonies, had earlier been restricted by weapons embargo measures by the United States. However, 'dual purpose equipment' as well as extensive financial resources to be extended to

Portugal were recommended. Another policy change concerned Rhodesia's chrome which was bought by the United States again – after having been officially boycotted. And hitherto increasingly restrictive relations with South Africa could be greatly relaxed.[23] The evident realist attitude of the document reasoned that to support black Africans in their liberation wars would only set the stage for further chaos from which Africa's radical and external communist interests could profit. The United States could exert more positive influence by remaining cautiously engaged rather than by a flamboyant and moralistic departure from the region. This rationale closely presaged the reasoning demonstrated in the Reagan era's contentious policy of 'Constructive Engagement'.[24]

Though *NSSM 39* never served as a blueprint for comprehensive action, it was a very revealing explication of the official United States thought processes of the Nixon-Ford years as they came to grips with the changing realities of the emerging Third World. This vast global segment would increasingly occupy the United States – whose previous attention had been overwhelmingly concentrated on great power posturing in the world's main theaters of contention. The document identified as one of the United States' objectives, 'to minimize the opportunities for the USSR and Communist China to exploit the racial issue in the region ... and to gain political influence with black governments and liberation movements'.[25] America's dilemma with regard to southern Africa was illustrated throughout the document as it surveyed our sparse economic relations with Africa, the need to maintain cohesion in NATO (which recognized the geostrategic value of the Portuguese Azores), the opportunities available to the Soviet Union in Africa's tumultuous events, and the need to value our established allies (perceived by Africans as usually being Caucasians), whether in Europe or Africa. Yet, *NSSM 39* injected a countervailing note, 'we sympathize with the aspirations of the Angolans and Mozambicans for self-determination'.[26] And in its introductory summary, the document refers to United States objectives which 'in some instances are conflicting and irreconcilable' – a sober self-assessment.[27]

These diplomatic dilemmas so well displayed in *NSSM 39* were not overcome in the ensuing years as the United States sought to remain actively committed in the affairs of the southern African region – and encountered great frustration in that attempt. However, the profound changes introduced in the region were not engineered by the United States but by other forces such as the internal revolution in Portugal; the shift of Ian Smith's Rhodesia to majority black rule; American state, municipal, and private non-governmental punitive sanctions against

South Africa; the implosion of the Soviet Union and the ensuing departure of Soviets and Cubans from Angola; a substantial brain drain and capital flight from South Africa's white community; the high birth rate and lack of jobs among black South Africans; Nelson Mandela's release from prison and his masterful assumption of leadership of South Africa's majority blacks; and the sober shift of key southern African black leaders from 'alleged' Marxism to overtly pro-Western and pragmatic ideological positions.

Essentially, the United States was mostly a half-committed, but well-meaning bystander, but in the end, a great beneficiary of rapidly changing events in southern Africa. In this region, the US engaged primarily in conservative reactive diplomacy. With hindsight, this minimalist and cautious course of diplomatic action may be justified, and it stands in stark contrast to the adventurist, if not aggressive, and costly approach taken by the Soviets and Cubans. In human terms, much blood was shed in Zimbabwe's war for majority rule and the short civil war after the fall of white domination. The prolonged insurgency in Mozambique surely rivaled the anarchy in Somalia a decade later. The long war that intertwined the domestic consolidative efforts of Angola's government with Namibia's independence struggle failed to generate a meaningful, permanent ally for the Soviets. Further regional instability was introduced by South Africa's military intervention in Lesotho and, of course, by the multipronged active opposition by the African National Congress in its escalating militant activities against the ruling white oligarchy. Yet throughout the region a direct superpower confrontation had been avoided. Where the United States had repeatedly been identified as the enemy of the 'liberationist' forces, the transition to diplomatic acceptance in the region developed very rapidly.

By contrast, explaining the history of Soviet behavior in southern Africa, as on the rest of the continent, has not yet been accomplished definitively. Western analysts offer a variety of explanations such as the Soviet search for key geostrategic allies in various regions of the globe; ideological proselytizing and revolutionary support; Cold War strategic opportunism; champion of the Third World sectors; the search for trading markets for their goods which were not of acceptable quality in the technologically advanced West; and the better-known explanations concerning Soviet designs for controlling strategic sea lanes around southern Africa, and the intention to take control over southern Africa's 'Persian Gulf of minerals' – and deny them to the West. Accordingly, Angola and Mozambique were the major regional targets of Soviet geostrategic interests. Regarding Zimbabwe, by not backing Robert Mugabe's

Shona-led ZANU movement in Zimbabwe's war for majority-based rule, the Soviets (mistakenly) backed Joshua Nkomo's competing minority Ndebele-based ZAPU guerrillas who represented only a 20 percent minority faction of Zimbabwe's population. And, in the case of South Africa, the Soviets equivocated in their dealings with the African National Congress despite the fraternal affiliation of South Africa's Communist Party with the ANC.

At the official level, the United States had focused on Soviet designs for support to African revolutionaries; many warned of the dangers of potential Soviet control of sea lanes; and the most popular explanation was the greatly exaggerated threat to take control of 'strategic minerals' in southern Africa. These reasons were often cited by American as well as South African conservative forces that inevitably argued in favor of tolerating continued white control of South Africa as long as the Cold War persisted. The first two of these perceptions were rarely examined in detail, while the minerals argument received wide business and congressional attention.[28] However, all three arguments had their detractors as well.

On rare occasions when revealing, but ideologically infused, information was forthcoming from the Soviets, explanations of their interests in Africa usually expressed support of fledgling socialist-tending states that needed external assistance against the overwhelming power of Western neocolonialist forces. Reflecting a long-held Soviet government position, S. I. Tyulpanov summarized it, 'The socialist-oriented countries' national strategy of struggle against neocolonialism envisages the establishment of strong political and economic ties with all Third World countries defending their sovereignty, regardless of their social orientation.'[29] This took them to Africa with but modest economic aid efforts, and political programs patterned on the Soviet experience as a model for development.[30] Indeed, early optimism led the Soviets to believe, in the words of Nikolai Kosukhin, 'in ... Africa ... one of the original features is the fact that objective possibilities have taken shape there in the present age for bypassing the capitalist path of development'.[31] However, a post-Cold War assessment by Vladimir I. Tikhomirov asserts that (in the case of southern Africa,) the region 'has never been of special interest in Soviet foreign policy'.[32] He intimated that since the early 1960s, the Soviet objective in Africa had been the support of national liberation movements. Thereafter the Soviet Union, like the United States, experienced few positive gains in Africa and began to reduce their involvement.

In the mid-1970s, with the collapse of Portugal's control over its African colonies, Angola's and Mozambique's weak new governments called on the Soviets to help them resist Western and South African support for internal opponents who had launched insurgencies in those fledgling states. Interestingly, the West used a similar justification for re-engaging in Africa, only they blamed it all on Soviet and Cuban involvement. This has led J. Gus Liebenow to remark, 'it is striking to consider the many parallels in attitudes and actions on the part of both major powers'.[33] And it may well validate the arguments of those who postulate that ideology does not drive the actions of great powers as does their realistic strategic assessments and their global reach. The lack of concisely stated and plausible explanations by both the United States and the Soviet Union regarding their actions in Africa had been a standard feature of the role of Africa in the Cold War.

With the assumption of power in the Soviet Union by Mikhail Gorbachev, the internal 'new thinking' about the Soviet Union's impending political revolution became evident in 1985. Explicitly, new directions in Soviet strategic and diplomatic approaches became apparent as the old, standard, liberationist explanations regarding Soviet support for Third World countries in the quest to expand the world's community of socialist states waned. Southern Africa may have yielded one of the earliest – if not the first – signs of the impending global disengagement by the Soviet Union. An early indication concerned Mozambique, which had initiated diplomatic steps to move firmly into the Soviet camp in 1977. Michael Clough notes that sometime in the early 1980s the Soviets informed Mozambique that it would 'not be invited to join the Council for Mutual Economic Assistance (COMECON)', and he reports that Mozambique was advised by the Soviets to seek economic assistance from the West.[34] If verified, to those who followed Soviet affairs, this was a very revealing development as it certainly reversed the tenor of traditional Soviet rhetoric concerning their efforts at increasing their leadership of the Third World.

However, there exists some doubt about this interpretation as Mozambican officials have, on occasion, denied that such a Mozambican appeal to join COMECON ever occurred.[35] However, rumors of this development had circulated in South Africa and was further corroborated by the signing of the Nkomati Accord between Mozambique and South Africa in 1984. That accord paved the way for accommodation between the two countries and called for the mutual ending of support to each other's governmental opponents.[36] United States diplomatic facilities in both countries had to readjust quickly. By the following year, the 'new thinking' in the Soviet Union also led to the ANC being advised by the Soviets that a peaceful accommodation should be pursued with

South Africa's minority government.[37] The Soviets would not be in a position to assist the ANC in a military victory. As the decade advanced, Mozambique soon showed signs of having been effectively abandoned by the Soviets, and came increasingly under South African influence, possibly to the point of having covertly tolerated the South African government's punitive strikes against the ANC cells in Maputo. One such foray never sufficiently explained was a car bomb attack against a highly placed ANC operative who miraculously survived. The fact that it took place on a holiday in 1988 only added to the suspicion as it would minimize collateral damage.

These were all major developments that emanated, not from United States policy initiatives, but mostly from tumultuous internal affairs in the Soviet Union whose consequences manifested themselves early in Soviet policy in southern Africa, and from which the United States subsequently benefited greatly. Of note is America's history of predominantly reactive policies to the Soviet Union's vacillating initiatives and commitments; the cautious and conservative American policies towards South Africa, Namibia, and, under President Clinton's presidency, Angola; and after the departure of the Soviets and Cubans from the continent, the rapid reversal from previous policies which had pragmatically, but unenthusiastically tolerated South Africa's continued white rule, to an untroubled current acceptance of black majority rule throughout the region. In effect, this validated America's contention that its policies had not been shaped primarily by nebulous, racial considerations but by Cold War imperatives which demanded that the defense of America's core security interests could not be compromised even in distant, peripheral war theaters. All worldwide conflicts had to be monitored, and many had to be effectively neutralized, however distant.

This constituted the essence of the Reagan era 'Constructive Engagement' policy towards southern Africa, though initially that policy was perceived to be more benign than it turned out to be. The policy amounted to a quiet but firm holding action that would balance off aggressive Soviet regional measures, but avoid a direct Cold War battlefront from opening in southern Africa. The policy met with moderate success, but reaped its major reward when the Soviets began voluntarily to abandon their distant commitments. Initially, there was loose talk of possibly a United States-Soviet Union initiative to work for peace jointly in Africa, but this amounted to little more than public face-saving rhetoric for the departing Soviets and was quickly abandoned. They couldn't keep their own house in order. Essentially, the United States attained its new diplomatic position in southern Africa by the grace of good fortune, and with little investment in costly efforts of its own.[38] This good fortune was further enhanced for the US in that the southern African states would intensify their external ties most naturally with Europe and the United States once peace returned to the region.

Southern Africa's Strategic Transition in US Policy

The rapid pace and profound impact of change in Cold War-related global developments caught most countries in the southern African region by surprise, while the United States had to readjust quickly to dramatic changes in this region. America's established regional interests, the departure of Cuban and Soviet military personnel from the region, the attainment of independence of Namibia, the progressive pacification of Mozambique, the release of Nelson Mandela and unbanning of the ANC, and the scheduling of elections in South Africa which would certainly lead to blacks assuming effective political power in that country, represented a kaleidoscope of far-reaching political transformations. Domestic pressures in the United States were prepared to again organize on behalf of South Africa's majority black population, but this was not now needed. Again, good fortune was manifest as the powerful Afrikaner Nationalist Party took the initiative to hold all-race elections – knowing that whites would yield power thereby and that apartheid would be buried. The United States was apprehensive, as the consequences of such radical turns could not be sufficiently predicted and could possibly destabilize the region.

The United States, however, gained from these developments as they greatly stabilized this important region in Africa, while South Africa gained from its political and diplomatic legitimization and overnight establishment and acceptance (although by some neighboring states, reluctantly) as the region's dominant leadership force. The coincidence of these revolutionary developments in South Africa with the 1991 Gulf War between the United States-led alliance and Iraq, quickly shifted America's strategic attention from southern to northern, and especially north-eastern, Africa. This action also marginalized most of sub-Saharan Africa, something Africans had expected might occur – but not so abruptly and so soon. President George Bush had demonstrated a greater degree of caution in dealing with Mandela[39], but the following Clinton administration's rapid ideological *rapprochement* with a new South Africa under Mandela's leadership left the rest of the southern Africa region to fend for itself against the newly legitimized 'colossus' in their midst. Publicly, South Africa's black-dominated neighbors had clamored for justice during the apartheid days in that country, but privately, they had been apprehensive about the anticipated operational implications of majority rule in South Africa. Above all, would South Africa's neighbors still be able to attract the generous foreign aid from United States and European sources, which previously was supplied in order to enable them to counter the financial cost of South Africa's apartheid policies, which had allegedly levied enormous direct and indirect financial costs on them?

Official United States policy towards southern Africa had traditionally focused heavily on South Africa as the focal point of weakness and turmoil in the region. In a rare policy statement by President Reagan regarding the region's major country, he had offered, 'The root cause of South Africa's disorder is apartheid ...', and 'apartheid must be dismantled.'[40] However, after insisting that apartheid is immoral, he was very emphatic that economic sanctions were also immoral, as they would hurt the black workers. This was a plausible argument also offered by some white South African anti-apartheid luminaries such as Nobel Prize winning writer Alan Paton. Reagan referred also to the hundreds of thousands of black workers from neighboring states that work as miners and very low-paid agricultural workers in South Africa, who would be hurt by such sanctions. However, Reagan appeared not be aware that South Africa's employment of laborers from neighboring countries had been designed as a means with which to encourage tolerant mutual relations between South Africa's white regime and the six contiguous black-run countries. Candidly, Reagan offered, 'In southern Africa, our national ideals and strategic interests come together.' Reagan also listed as important considerations the usual, well-worn arguments concerning the strategic sea lanes and access to the vital, or strategic, minerals which needed protection. He cautioned, that any mistakes in policy, and 'the Soviet Union would be the main beneficiary'.

These were old arguments that had been touted by the political right in the United States as well as in South Africa, but were never convincingly supported with credible data. Did anyone really believe that the Soviets had the requisite manpower resources with which actually to administer grandiose ambitions such as physically taking over control of the distant sea lanes? And, could the Soviets effectively occupy and control the vast southern Africa region in order to curtail the exports of raw materials which comprises the major source of export earnings and provides a large portion of jobs and incomes for all sectors in southern African societies? Besides South Africa, the economies of Namibia, Botswana, Zimbabwe, and Zambia also relied on mineral exports for the stabilization and growth of their respective economies. Nor did the Soviets need these resources for their own domestic use as they had sufficient reserves of these minerals. Still others had argued that, indeed, if blacks came to power in South Africa, the new government's regional policies might be more damaging to neighboring states, as black-run South Africa would restrict labor inflows from these states in order to reserve the jobs for its own black workers – of whose available workforce, close to half were unemployed.[41]

Under President Bush the annual White House policy statement *National Security Strategy of the United States* formalized a new theme, but in steadily more abbreviated references to Africa. Accordingly, in the absence of the Cold War, Africa could now progress on its own path towards development unimpeded by external conflicting interests. The United States would actively support black and white leaders in South Africa 'to move that country into a democratic, constitutional, post-apartheid era'.[42] However, abruptly changing America's policies towards South Africa posed a dilemma as an assessment of Mandela's full powers and influence among the untested black majority had not yet occurred, nor was it certain initially that the ANC's suspected proclivities towards Marxism would be effectively abandoned.[43] In fact, Joe Slovo, the head of South Africa's newly legitimized Communist Party, had held a ministerial position in Mandela's initial administration. Hence to some, there was a valid basis for America's initial concern. But with the Soviet and Cuban departure from Angola, and with the ANC's rapid turn towards ideological and organizational moderation, it was not difficult to sense the onset of America's strategic downgrading of the entire region, and the rapid shift of United States attention and action to north-eastern Africa and the changes in this region brought about by the Gulf War of 1991, and by the Somali crisis.

Another development in America's posture was the shift of interest in southern Africa from official United States governmental institutions to the media and the private sector, as non-governmental economic interests undertook a measure of the profit potential of southern Africa – meaning, of course, South Africa. That country was quickly labeled an 'emerging market' by the United States government, a policy in which the private commercial sector is encouraged to expand its interests, abetted by active United States government assistance. Within the first years of South Africa's new commercial status, American companies began to trickle back but not in the quantities evident before the effective United States disinvestment campaigns of the previous decade, and in only modest undertakings geared towards establishing a physical presence, or observation vantage point, but not large-sized ventures.

A black-governed South Africa, together with the pacified southern Africa region, offers some promising physical ingredients for the development of a substantial integrated market by Third World standards, but the transition towards its actualization would be greatly conditioned by the influence of the generally declining economic fortunes of the rest of the continent. A repeated complaint by foreign corporations is that Africa does have the potential of developing modest-sized, regional markets, but while investors might earn 10 percent of their overseas profits in Africa, they must expend half of their managerial time and effort in its pursuit. Also, mature analysts understand that it will take several years – if not decades – after such a profound political transition as has occurred in South Africa, for the internal clashing and competitive socio-economic forces to emerge and yield a lasting and stable profile of new de facto power foci. The African continent is replete with examples of internal tensions escalating to outright civil wars long after revolutionary social changes had been introduced at independence and as the new polities pursue their territorial and national consolidation. Reportedly, in the decade before the formal termination of apartheid, an enormous amount of white-owned South African capital fled the country for safer overseas havens, while substantial numbers of young, white university graduates left the country for more promising long-term career opportunities elsewhere. The United States was one of the leading destinations for South African emigrants. In South Africa, during the late 1980s, rumors were rife that high-ranking white officials were receiving part of their salaries paid into overseas bank accounts. In effect, South Africa today may be expecting that the foreign investor community, hopefully led by Americans, will develop the region – and face a substantial amount of financial risk during the present transitional stage of southern Africa.

The Economist reported that 1996 was a very favorable year for economic growth for most of the members of the Southern African Development Community (SADC), but this should not mask the annual growth vacillations of individual states and the need for expanding costly infrastructural links. The report also recognized the inherent limitations of a potential combined regional market and its domination by South Africa.[44] However, in light of the recent political democratization and the stabilization and expansion of markets in Latin America and East Asia, America's private sector interests in Africa will be very limited, while humanitarian concerns will remain the prerogative of only official attention. Looming over this latter prospect will be increasingly more instances of 'donor fatigue', as well as reluctance of direct intervention in ongoing internal turmoil – the consequence of America's unfortunate experience in Somalia.

The Clinton Administration had the good fortune of inheriting a southern Africa region well along in its pacification and historical transition. But then again, there were no realistic options for the region's states beyond reigniting traditional relations with the British Commonwealth, and with the United States, whose Democratic Party had a historically more sympathetic attitude towards black Africa's plight. In summary form, the Clinton White House expressed its initial foreign policy posture in its document *A National Security Strategy of Engagement and Enlargement*.[45] The President warned of new types of security challenges, but also of new opportunities for democracy, peace, and prosperity. His administration identified as the most significant indicators of the new era the declining prospects of nuclear annihilation, the expectation of progress towards peace between Israel and the PLO, and notably, it lauded the 1994 multiracial election in South Africa – that resulted in Mandela's presidency. The United States would remain committed to ensuring that democracy takes root in South Africa, which is to 'foster a new era of prosperity and stability for all peoples in the region'.[46] More specifically, the following year's statement added that during a visit to Washington by Mandela, 'we announced formation of a bilateral commission to foster new cooperation between our nations'.[47]

In another White House statement, reference was made to the United States role 'to help bring an end to two decades of civil war and to promote national reconciliation' – in Mozambique and Angola.[48] And, under the category of 'Promoting Democracy', the annual national security statement referred to South Africa, whose entry into 'the camp of market democracies may influence the future direction of an entire region; South Africa now holds that potential with regard to sub-Saharan Africa'.[49] Clearly, the Clinton Administration expressed only optimism about South Africa's potentially future activist regional role which, however, lacked a qualifier that acknowledged the need for sensitivity to the justified apprehensions about such a powerful and potential hegemon in the midst of comparatively very weak states. On a continent wracked with armed conflicts, the mere accession to power of a majority, black-led regime in South Africa did not automatically guarantee South Africa's future regionally fraternal behavior.

The central thrust of these current but greatly abbreviated American perceptions of southern Africa justifies the views of skeptics who identify two clear trends: America's perceived tolerance regarding the realities of the debilitating marginalization of much of Africa, and placing a perhaps unwarranted responsibility on South Africa to fill the power and leadership void in the greater region. Africa can simply consume too much diplomatic energy while producing little, if any return on externally derived financial and diplomatic investments. South Africa should be encouraged to develop its economic engine in order to service the commercial needs of its region and possibly, much of the sub-continent. However, as in the case of Nigeria in West Africa, the nurturing of an indigenous regional leader may in the end amount to little more than abetting the emergence of a regionally-based 'neocolonialist power' – as it will be perceived by a weak neighbor. America's downgrading of its interests in the numerous sparsely populated countries in Africa was amply illustrated when Clinton announced the closure of the United States embassy in Comoros, citing American budgetary pressures as the cause.[50] The moribund economy of Comoros, which relies on the export of only a few spices and some perfume ingredients, has been effectively marginalized, and may be a harbinger of what to expect in the future of United States relations with numerous other such African states.

A new direction of America's policy towards Africa may be detected in the May 1997 version of the national security strategy titled *A National Security Strategy for a New Century*. The usual positive introduction is offered, but then a major thrust focuses on the raging conflicts, including references to state-sponsored terrorism, narcotics trafficking, international crime, environmental damage and disease in Africa. Beyond repeating what is already well known, the document reveals not a humanitarian perception as much as it places Africa into a context that describes our standard perceptions of the entire Third World. Africa, and specifically southern Africa, will not be the beneficiary of a costly partnership and innovative relations with the United States. Rather, the United States will monitor especially the northern and north-eastern portion of the continent in its potential security context of providing advantage for extraregionally-derived destabilizing activities that may serve the interests of neighboring aggressors in the Middle East and Mediterranean regions. In this geopolitical context, we may reasonably expect a reduced American official valuation of the distant and isolated southern Africa region, although the international commercial community may find advantage in helping to develop an expanded and integrated regional economic network. For this to materialize, it will be essential that South Africa advances in a climate of domestic stability, brought about by egalitarian economic progress for all social sectors – a formidable challenge. Where the United States had reacted to Soviet-allied initiatives during the Cold War, with respect to southern Africa, in the future the United States will likely be reacting to South Africa's domestic and regional agendas.

Notes

1. Robert Lloyd George (1994) (somewhat optimistically) lists nine emerging markets in sub-Saharan Africa of which five are identified in southern Africa. *The Handbook of Emerging Markets* (Chicago: Probus).
2. Michael Clough (1992) *Free at Last? US Policy Toward Africa and the End of the Cold War* (New York: Council on Foreign Relations Press), 21.
3. Christopher Coker (November 1994) 'Western Intervention in the Third World,' *The Officer*, 34.
4. Mozambique's President Samora Machel allegedly advised Zimbabwe's President Robert Mugabe before majority rule commenced, 'Keep your whites.' Quoted by Keith B. Richards (1997) *Out of America* (New York: Basic Books), 215.
5. Karl P. Magyar (6 October 1994) 'Against Heavy Odds in South Africa', *Washington Times*.
6. Ieuan LL. Griffiths (1995) *The African Inheritance* (London: Routledge), 120–1.
7. Quoted in Karl P. Magyar, 'South Africa's political economy: Perspectives on the country as a fragile economic arrangement', in Albert Venter (ed.) (1989) *South African Government and Politics* (Johannesburg: Southern Publishers), 228–9.
8. James H. Polhemus refers to the hostage theme in 'South Africa's Hostages? The BLS States in a Changing southern Africa', in Paul B. Rich (ed.) (1994) *The Dynamics of Change in southern Africa* (New York: St. Martin's Press), Ch. 11. He also refers to Jack Halpern's book *South Africa's Hostages: Basutoland, Bechuanaland and Swaziland.*
9. The controversial historical accounts of the arrival of various peoples in southern Africa is summarized by Kevin Shillington (1989) *History of Africa* (New York: St. Martin's Press), Ch. 15.
10. Ieuan LL. Griffiths (1993) *The Atlas of African Affairs* (London: Routledge, 2nd edn), 48.
11. The little-known historical and contemporary legacy of this politically significant community to American audiences is presented in Anthony J. Arkin, Karl P. Magyar, and Gerald J. Pillay (eds) (1989) *The Indian South Africans: A Contemporary Profile* (Pinetown, South Africa: Owen Burgess).
12. The 'prolonged nature' of these wars is reviewed in Karl P. Magyar and Constantine P. Danopoulos (1994) *Prolonged Wars: A Post-Nuclear Challenge* (Maxwell AFB, AL: Air University Press, 1994).
13. Paul Y. Hammond (1969) *The Cold War Years: American Foreign Policy Since 1945* (New York: Harcourt, Brace and World), 177.
14. The precise sequence of Mobutu's rule is disputed. Griffiths (*Atlas of African Affairs*) lists Mobutu's assumption to power as 1960 and 1964, 84–5.
15. Clough (1992), 79–83.

16. Immanuel Wallerstein, 'Africa, the United States, and the World Economy: The Historical Bases of American Policy', in Frederick S. Arkhurst (ed.) (1975) *US Policy Toward Africa* (New York: Praeger), 30.

17. 'Clinton Supports Comoros "Territorial Integrity"', *Africa News ONLINE* (http://www.africanews.org/usafrica/stories/19971118_feat1.html, 11/27/97), 2.

18. Paul L. Moorcraft (1990) *African Nemesis: War and Revolution in southern Africa* (London: Brassey's), 63.

19. Gerald Bender, 'American Policy toward Angola: A History of Linkage', in Gerald J. Bender *et al.* (eds) (1985) *African Crisis Areas and US Policy* (Berkeley: University of California Press), 110.

20. On an inspection tour of the war zone of northern Namibia in 1988, I attended a military briefing at Katima Mulilo, a military base in Eastern Caprivi. During one briefing, an aging officer produced a map with all of Africa – except for South Africa and Namibia – painted red and with numerous arrows, representing the 'communists' pointing at South Africa. This slide expressed most graphically the Rooigevaar that was at the base of the whites' political fears.

21. A quick review of United States–Angolan relations is offered by Tetteh Hormeku (1997) 'US Intervention in Africa: Through Angolan Eyes', *African Agenda*, no. 12/13. 11. See also the critical commentary on Reagan's false perceptions regarding southern Africa, Vincent Khapoya (1994) *The African Experience: An Introduction* (Englewood Cliffs, NJ: Prentice Hall), 225–8.

22. See the comments by Colin Legum (1988) regarding the Soviets, Cubans, Chinese, and Americans in southern Africa, *The Battlefronts of southern Africa* (New York: Africana), Introduction.

23. Moorcraft (1990), 31. Also, Mohamed A. El-Khawas and Barry Cohen (eds) (1976) *The Kissinger Study of southern Africa* (Westport, CT: Lawrence Hill), 45.

24. A collection of criticisms of Constructive Engagement is offered by Phyllis Johnson and David Martin (eds) (1988) *Frontline Southern Africa: Destructive Engagement* (New York: Four Walls Eight Windows).

25. Ibid., 82.

26. Ibid., 97.

27. Ibid., 82.

28. An explicit, authoritative identification of US policy along these lines is offered by then-Senator George McGovern (1979) *Impressions of southern Africa*, A Report to the Committee on Foreign Relations, US Senate (Washington, DC: US GPO), 5–7.

29. S. I. Tyulpanov, 'The National Strategy of Socialist-Oriented Countries', in E. A. Tarabrin (1978) *Neocolonialism and Africa in the 1970s* (Moscow: Progress), 320.

30. Nikolai Kosukhin (1985) *Revolutionary Democracy in Africa: Its Ideology and Policy* (Moscow: Progress), 165.

31. Ibid., 97.

32. Vladimir I. Tikhomirov, 'Contemporary politics in South Africa and the Soviet policy towards southern Africa', in Anatoly Gromyko and John Kane-Berman (eds) (1991) *The Moscow Papers. The USSR and South Africa: Similarities, Problems and Opportunities* (Johannesburg: South African Institute of Race Relations), 9.

33. J. Gus Liebenow (1986) *African Politics: Crises and Challenges* (Bloomington: Indiana University Press), 267.
34. Clough (1992), 12.
35. The denial was expressed to the author in several interviews in 1989 and 1990.
36. Gwyneth Williams and Brian Hackland (1988) *The Dictionary of Contemporary Politics of southern Africa* (New York: Macmillan), 185.
37. Tikhomirov (1991), 12.
38. Karl P. Magyar, 'Namibia: The International Politics of Transition', in Marion E. Doro (ed.) (1992) *Africa Contemporary Record, 1988–1989* (New York: Africana), B-625–6.
39. See Karl P. Magyar, 'Sub-Saharan Africa: Political Marginalization and Strategic Realignment', in James E. Winkates, J. Richard Walsh, Joseph M. Scolnick, Jr, (eds) (1994) *US Foreign Policy in Transition* (Chicago: Nelson-Hall), 242–52.
40. President R. Reagan (22 July 1986) 'Ending Apartheid in South Africa', *Current Policy No. 853* (US Dept of State).
41. Always a controversial topic, Ahmad A. H. M. Aly (1997) quotes 1995 data from *The Economist*, and *World Development Report*, which reported South Africa's unemployment rates ranging from 33 to 50 percent. 'Post-apartheid South Africa: The implications for regional cooperation in Africa', *Africa Insight*, 27 (1), 27. Universally accepted figures on South Africa's unemployment rates have always been a controversial topic as more meaningful data should reflect rates of gross underemployment due to many blacks being registered as workers, but at part-time or informal sector jobs which pay very little. Karl P. Magyar (1989), 'South Africa's political economy', 216–17.
42. The White House (August 1991) *National Security Strategy of the United States*, 10.
43. Karl P. Magyar, 'Sub-Saharan Africa: Political Marginalization and Strategic Realignment', in James E. Winkates, Richard Walsh and Joseph M. Scolnick (eds) (1994) *US Foreign Policy in Transition* (Chicago: Nelson-Hall), 244–52.
44. 'An African success story' (14 June 1997) (n.a.), *The Economist*, 47.
45. The White House (July 1994) *A National Security Strategy of Engagement and Enlargement*.
46. Ibid., 26.
47. The White House (February 1994) *A National Security Strategy of Engagement and Enlargement*, 31.
48. Ibid., 5.
49. The White House (July 1996) *A National Security Strategy of Engagement and Enlargement*, 33.
50. 'Clinton Supports Comoros "Territorial Integrity" ' (18 November 1997) *Africa News Online*, www.africanews.org/usafrica/stories/19971118_feat1.html.

Conclusion
Africa's Political and Strategic Marginalization: A United States Perspective

Karl P. Magyar

> When you go around the world and look at US national interests, Africa as a continent has to be near the bottom of the list.
>
> Mr Lawrence Eagleburger
> US Secretary of State[1]

Africa's Contemporary Conflict Context

Our analyses demonstrate that Africa remains as volatile as ever, much as Pliny observed two millennia ago, and that the conflictual character of Africa dominates the investigation of all other social affairs of the continent. Many of Africa's current tumultuous developments have been initiated and shaped by the Cold War, especially by its tendency to pressure vulnerable Third World states into two opposing camps (e.g. Angola). Other conflicts, as well as social problems associated with modernization efforts, may be traced to the fundamental changes brought about by external penetration of Africa's traditional societies and subsequent irreversible social disruptions encountered during the long era of European coastal trade, then effective colonial rule (e.g. Gold Coast). And numerous developmental problems resulting in conflicts emerged during the early volatile consolidative phase following independence (e.g. Nigeria). Of course, it may be argued that these sources of conflicts cannot be considered as mutually exclusive and therefore the origins of Africa's severe incidences of instability will represent a mix of causes.

The craft practiced by the historian in search of causes or origins of social phenomena remains an inherently subjective exercise. The analysts represented in this volume no doubt agree that most modern conflicts and other debilitating socio-economic developments in Africa were not directly the product of only the colonial period, nor exclusively of the Cold War era. For example, the numerous junior military officer-led, *coups d'état* reflect problems associated with standard nation-building internal processes. Setting aside the question of ultimate causation, we may safely expect that Africa's instability will most likely continue as such post-independence domestic volatility is generally in evidence in most new political societies embarking on tortuous internal consolidation and legitimization processes. In this, Africa is not immune. In Africa's recent history, certainly the externally based Cold War loomed large as an influential political factor that partially shaped the volatile character of the modern African state. However, today indigenous-derived forces may be identified as major contributors to the controversies regarding the continent in its search for relevance of not merely its geographic or mineral resources, but also its human cargo, whose value over the last few centuries was established more often by external commercial or strategic calculations than by humanistic considerations.

Development and Evolution

Today's problematic developmental environment in Africa stems from the attempts by these new states to transit from the traditional to modern forms of societal organization in a greatly compressed time frame.[2] The intervening transitional phase has universally been the most conflictual period in the evolution of states – African or non-African – as the internally derived pressures of rising expectations, urbanization, detribalization, Westernization of education, cultural adaptation, sectoral competition, dislocation, and powerful external, partisan-allied, interventionist forces converge to impede the peaceful evolution towards legitimate modernizing popular political order in these fragile states. During the Cold War days Africa's fate was determined significantly by external strategic-centered, partisan interests. Yet in this new era we may rightly ask if the continent's fate will reside even more than previously in the hands of external interests – this time, commercial.[3] At issue is not only the powerful problem of trade and investment concerns in today's highly competitive international commercial environment, but also the sheer political power that the world's leading multinational corporations bring to the bargaining tables of the Third World.

Not well known to the external public is the objective, but revealing, demographic data characterizing Africa. Morally, the sovereignty of the present African states is widely defended, especially their membership in the United Nations. Yet few realize that the median sub-Saharan African state contains only about 7 million people, and that 37 of Africa's 51 countries are inhabited by less than 10 million people each. Indeed, 17 of these states are home to under two million inhabitants each. These data must be taken into account by those who would sincerely advise and support standard socio-economic measures for the development of the continent. Most of Africa's states lack realistic prospects for respectable, diversified development. For this, and other historico-cultural reasons, Africa's states ought to be examined within the structure of a regionally integrative context which surely lies in Africa's future. The present market sizes of Africa's states is simply too small to deal productively with many foreign commercial interests. Investing to ensure their viability will no doubt encounter inherent limits to their developmental prospects. Within their inappropriate Westphalian nation-state context, the fragile and artificial nature of Africa's modern states will subject the continent to a variety of highly competitive external public and private forces against which Africa will not prevail. As presently constituted, at the level of diplomatic relations, the typical African state has struggled for sheer existential relevance. Except for the physical advantages of minerals and geostrategic facilities – the latter of which has greatly atrophied in the present age of sophisticated airpower – most African states bring little advantage to an alliance. Their visibility in the United Nations' General Assembly today does not carry the weight it did in the midst of the Cold War. Active involvement in the future affairs of the continent by external interests, such as the OECD countries of the First World, will reflect primarily economic and humanitarian matters. In pursuit of humanitarian objectives, external publicly supported efforts will encounter limits if the accumulated investments in these periodic charitable ventures do not result in progressive and lasting improvements. By contrast, the representatives of the international private sector may be expected to make more frequent developmental exploratory forays than will the public sector. The mission of the private sector is the pursuit of profits, with any humanitarian considerations being but a calculatable impediment, exacerbating the cost of doing business in the Third World that does not successfully adhere to the orthodox terms of the international market place.

America's private sector activities in Africa have not posed significant problems or scandalous behavior on the scale of perennial, dubious commercial ventures by Europe- or Orient-originated efforts. Currently America's trade with Africa totals only 2 percent of our worldwide trade, with two-thirds of our imports from Africa constituting petroleum, mostly Nigeria's. This low volume of economic transactions no doubt explains the generally tranquil relationship between Africa and the United States. However, questions have periodically arisen with respect to occasional controversial diplomatic policies emanating at the official level. Prominent examples include America's early misadventures in the Congo-Katanga-Lumumba embroglio; the problematic support of Portugal's ill-fated attempts to retain its African colonies; the Reagan Administration's Cold War pragmatism regarding the southern Africa region; United States Ambassador to Kenya Smith Hempstone's public undiplomatic pronouncements against Kenya's President Moi; frustrations encountered in the attempt to deal with unethical business practices when they do emerge – especially bribery issues in Nigeria; oil purchases from Angola's government whose respectably capable insurgents the US supported against that same government; our diplomatic stand-off with Nigeria's successive military regimes yet remaining Nigeria's best customer of their high-quality oil; begrudgingly commending Nigeria's effective initiative and leadership of ECOMOG's peacekeeping operations in Liberia; propping up at very great economic expense Liberia's self-selected President Doe – all to no avail; and supporting at great political expense Zaire's President Mobutu – held by many as having been Africa's worst corruption offender. This list is but a start.

However, the greatest skepticism regarding US activities in Africa since 1960 has, with good reason, been expressed towards our vacillative official relations with South Africa. This was well illustrated with the changing fortunes of the disinvestment campaign against South Africa which saw the relative political success of US state or lower-level official policies while at the national level a much more conservative policy of 'Constructive Engagement' was formulated and implemented. For much of the American public the issue was a moralistic black-and-white dilemma, while the United States government was guided by the principle of not wishing to let events deteriorate to an intensive, direct or surrogate confrontation with any ambitious Soviet Union and Cuban-led initiative. In fact, most of the punitive and effective economic measures against South Africa were leveled by private groups or officials at the municipal or state governmental levels. As a portion of America's total investments during the disinvestment campaign, United States sums involved were not great and concerned about 150 small and only 4 substantial-sized companies with representation in South Africa. Far more damaging to South Africa than the recall of investment capital was the isolation of that country from the dynamic expansion of the global economy and that country's mounting inaccessibility to new technology.

During the height of the Cold War's competitive era, the frustrations the United States encountered in Africa concerned dealing with the usual two regional powers which could not simply be ignored: South Africa and Nigeria. These two countries may realistically be expected to retain their controversial roles within their respective regions. During the last several decades, both have ensured that economic relations with the United States must remain in place; indeed, expanding mutual commercial ties will remain a desirable objective. However, both countries have also demonstrated a proclivity towards certain internal political, and external diplomatic actions, which the United States judges to be objectionable.

America's Conflict Intervention

America's official military visibility in Africa was most prominent at the Cold War's strategic context and it is in this sphere that we may expect the policy debate regarding Africa to continue. Future American interventionist military combat forays in Africa will concern primarily 'military operations other than war' (MOOTW) expeditions. Armed military operations will be kept at the low intensity level. After Somalia, only remotely should the United States expect a major conventional level operation in Africa, but should it occur, it would be reactive to destabilizing initiatives undertaken by others.

Given the nature of Africa's contemporary armed conflicts which are overwhelmingly civil or secessionist wars, America's recent, low-level military operations in Africa concerned the consequences of political chaos (Rwanda), massive social dislocation (Liberia), evacuation of US personnel (Zaire), severe humanitarian problems (Somalia), and large-scale, transnational, politically induced migrations deriving from lethal internal wars (Rwandans into Zaire and Tanzania). Except in Somalia, the United States has avoided combat in Africa as our intervention is motivated largely by humanitarian objectives, or by the need to evacuate Americans – and others – when order breaks down in various countries. If such struggles are not soon ended or contained, they may be expected to spill into neighboring lands as was the case with Sudan's, Liberia's and Rwanda's recent domestic conflicts, or these conflicts may soon be co-opted by neighboring or even distant interventionists. This latter tendency was illustrated in the conflicts in Ethiopia, Zaire, Chad, Sudan, Angola, and Mozambique. Where co-optation was successful, the armed conflicts quickly transformed into 'Prolonged Wars' – with their attendant destabilizing character for the wider region.[4] Most states throughout the world experienced similar domestic armed conflicts in their own early days as independent states. But in Africa this phenomenon is greatly exacerbated in that many contiguous states are undergoing this transitional nation-building process simultaneously; almost all are extremely poor and ill-equipped to address the lethal and unrealistic rising expectations of their inhabitants; and the appearance as independent states occurred in a highly competitive and interventionist-prone, international environment – in the throes of the Cold War. Accordingly, defining United States objectives in terms of only interventionist expeditions today would be a formidable challenge.

Monitoring the Sources of Africa's Contemporary Instabilities

Today, not all of Africa's socio-economic or security problems are the direct legacy of the divisive and hostile Cold War's competition for allegiance. Post-Cold War era domestic and global developments are increasingly making themselves felt in Africa. A review of these currently emerging problems suggest situations that will confront America's policy towards Africa in the context of the new era. Major armed, hostile incidences in Africa will continue but where necessary, if addressed, few of them will require a United States military response higher than MOOTW-level. However, Africa's conflicts must all be monitored closely in order to obviate the chance that they could deteriorate, spill into neighboring states, and pose new security challenges at the regional level, at which competing external forces may introduce their own interests and inflict long-term damage.

Current African developments of interest to the external analyst may be classified into three categories:

1. Ongoing negative or deteriorating developments that have characterized the last three decades, but emanate now from internal, domestic, and not from external, Cold War-related causes. These developments concern the problems associated with the pursuit of legitimate political orders; frustrations encountered in the attempts at national sociopolitical integration; the disruptions ensuing from the rural exodus to the cities; hunger, refugees, large migrations, poverty, disease, and recently, the highly disruptive social effects of AIDS. However, the single, greatest source of frustration is rising public expectations.

In the economic realm, ongoing problems will include the deterioration of the largely inherited infrastructural facilities; inability to industrialize and diversify profit-producing activities; faltering agricultural outputs; economic dislocation; uneven development; unemployment and under-employment; widening wealth divergence of top and bottom economic classes; continued reliance on primary exports; deteriorating terms of trade; debts, and shortage of domestic capital; inherently small market sizes; and the insufficiency of managerial development.

In the security realm, although certain wars have ended or are diminishing in scale and intensity, Africa's well-established history of armed conflicts show no signs of cessation. These include transnational wars, such as that between Somalia and Ethiopia; and especially civil wars, such as those in Sudan, Mozambique, Angola, Sierra Leone, Chad, Rwanda, and Uganda – all of which have experienced some periods of external intervention, but which also saw periods of domestic factional animosity – Rwanda being the most evident current illustration.

Africa's ongoing history of military *coups* and military rule further retard the development processes. Rarely does military-induced stabilization result in notable economic growth. Concerning America's major strategic interests in Africa, the continent shows some signs of stabilization but now significant instability has penetrated into the Sahelian belt ranging from Senegal to Somalia where a variety of escalating armed conflicts are being waged. On the positive side, except for the Rwanda/Burundi region, few large-scale transnational hostilities are being experienced in sub-Saharan Africa. However, expecting lasting resolutions of many of the internal conflicts reflects general Western naïveté regarding Africa's conflict culture especially in their Third World context.[5] Major wars could easily start again.

2. Developments emanating from the rapidly expanding, highly competitive, international socio-ideological environment. The impact of new global political forces is difficult to assess as they are emerging at the present time. What may well be the most important international ideological development from the point of view of Africans, is the bankruptcy of the orthodox socialist model as previously epitomized and propagated by the Soviet Union. Of course, it may be argued that Africans never adopted that model in its evolutionary European context. Others argue that Marx's requirement of the Historical Dialectic, did not address the pre-industrial conditions prevailing throughout Africa. And we are aware that generally, the socialist states have been no more successful at generating significant unqualified aggregate growth than have the free market oriented states.

Evidence is emerging of the broad discontent within Africa's ruling circles regarding economic structures which may be characterized by the standard socialist attributes such as public ownership of enterprises, centralized planning, egalitarianist social policies, heavy industrialization, redistributionist measures, financing ambitious social programs, etc. In effect, the Soviet-initiated, socialist ideological rug, along with the former Soviet Union's sparse economic aid programs, had been pulled out from under their sparse admirers in Africa. However, ample examples also exist of the Africans' rejection of America's oversimplified ideological mantra concerning human rights, democracy, competitive elections, free markets, free press, and peaceful conflict resolution.

Seen from another perspective, the late Soviet Union's well-established and substantial, security commitments to countries such as Angola and Ethiopia, and previously to others such as Ghana, Mali, Central African Republic, Somalia, Mozambique, Benin, etc., have rapidly come to an end. This also left the Ethiopian Marxist revolution in complete limbo with Eritrea gaining its liberation from Ethiopia's 'colonial' rule. In Angola, the Soviets had correctly advised the governing MPLA that Angola's insurgent force UNITA, could not be defeated militarily. That conflict has to this day not been conclusively resolved. Cuba retrenched on schedule due mostly to the lack of Soviet financial and military support, and to South Africa's radical turn in domestic affairs. Since ascending to power in Namibia, SWAPO, the ruling authority led by President Sam Nujoma, has had to rein in its traditionally more socialist inclination and has agreed to stable, cooperative terms with the new, majority-run South Africa. Mozambique has had to make peace with South Africa at Nkomati as early as 1984, as the Soviets were in no position to sponsor the FRELIMO government in its aspiration to take a decisive offensive against their South Africa-supported MNR insurgents. And in South Africa, the white government engineered a surprisingly peaceful and rapid power conversion to the majority black sector. In all these dynamic developments, the United States played mostly a moderately supporting or mediative role. However, in the process the United States revealed the prospect of tolerating a strong, leading state in each of Africa's five geographic sub-regions. But except for South Africa in southern Africa, anointing such other leaders has amounted to an exercise in frustration as regionally dominant states can quickly graduate to hegemonic ambitions, or they are part of the problem, and not the solution (e.g. Nigeria and Congo.)

With most of Africa's old guard having left office, and in view of the failure by Africans broadly to advance their own development, Eastern

Europe's democratization has also made its influence felt by relegating Africa to an even more obscure diplomatic position than it has previously held. It may well be the case that Africans correctly anticipated their diminishing value, and therefore Africans exaggerated their new respect for democratizing their domestic order in conformity especially with America's political demands. Perhaps this would attract America's generosity to their desperate plight. The most evident manifestation of political transformation is the move towards tightly controlled multiparty structures in several countries which coincides with the concomitant demise of military authority where they may still be in effective power. Besides Zambia and Kenya, democratization pressures have also been felt in the francophone states such as Benin, Gabon, Cameroon, Ivory Coast, Congo-Brazzaville, and the Democratic Republic of the Congo – some of which were already the traditionally more stable states in Africa. The more radical opposition groups have in turn been influenced to moderate their programs before assuming power. These include Ethiopia's new government, SWAPO in Namibia, FRELIMO in Mozambique, the MPLA in Angola, and the ANC in South Africa. All promised more militant programs than they actually produced once in power.

But, in the early 1990s, with global attention focused on events in Eastern Europe, on the evaporation of the Soviet Union, and on conflicts of greater global consequence to the developed world such as the war over Kuwait, Africans across the continent expressed fears of losing the battle against marginalization. The continent is no longer a location of Soviet-American rivalry; it remains politically unstable; and remains economically greatly underdeveloped. Its portrayal in the media is often as a large, hopeless social entity. Ironically, Africa may become the dark continent once again. In essence, Africa may be feeling the negative effects of post-containment global changes, but it is of little consequence or overtly articulated concern to the developed external world.

3.   Recent developments which may portend a reassertion of Africa's regional and global strategic role. The United States had viewed the Soviet Union and its Cold War-allied states as determining Africa's major regional security role, based on a combination of physical, geographic, and demographic factors. This had led to Africa being selectively divided into flexible, if not pragmatic, spheres of interests, as was demonstrated when the two superpowers unceremoniously switched their allegiance from among Ethiopia and Somalia. Such political posturing would have been remarkable elsewhere in the world, but in Africa it illustrated the largely unprincipled Cold War logic that would also

lead to the general perception of Africa as a marginalized continent in the new era. However, the diminution of those Cold War divisions in Africa does not mean the elimination of all security-related concerns from the continent. New, external forces have quickly filled the void. A greatly under-reported example of new, externally originated forces illustrates this point.

During the prolonged war between Iran and Iraq in the 1980s, the two countries had competed for a variety of diplomatic advances in Africa, in opposition to each other. In general, Iraq pursued expanded relations with the solidly Muslim states of North Africa and the Sahelian belt from Mauritania to the Horn, while Iran had embarked on an ambitious diplomatic offensive in pursuit of allies in sub-Saharan Africa. Iran had soon doubled her diplomatic missions in Africa. Countries which had been targeted for Iranian aid, exchanges, scholarships, and intensified political relations, included Nigeria, Sierra Leone, Zambia, Ghana, Tanzania, and Kenya. Of interest were the promises of generosity made to Namibia after its separation from South Africa, and an invitation for a visit to Iran to Nelson Mandela, once freed. Muslims in South Africa had also been actively coveted by Tehran. Iran's relations also strengthened with Senegal, which had been locked in a long confrontation with Iraq's firm ally, Mauritania. Iran's sub-Saharan Africa offensive appeared to concern primarily anti-Iraq diplomatic objectives, building economic relations, and the expansion of Shia Islam.

Iraq soon matched Iran's efforts, which underlined a new diplomatic contest over Africa during the height of their war. Iraq's more aggressive political initiatives envisioned the use of Africa as a broadened battleground for the Middle East battle zone. In Africa, Iraq had discarded her concern with ideological and religious pursuits and, instead, focused her entire offensive pragmatically on the political and strategic realm. Though unconfirmed, media sources reported that Iraq had intended to install a test site for her ballistic missiles in Mauritania, and to pursue basing facilities there for elements of the Polisario Front in retaliation for Morocco having sent troops to Saudi Arabia in the confrontation over Kuwait. And wartime rumors identified Sudan as possibly allowing Iraq to emplace Scud-B surface-to-surface missiles on her northern border with Egypt – a mere 120 miles from the Aswan Dam. Though lacking subsequent validation, such rumors pointed to future prospects of Africa's strategic value in widening any wars in the Middle East. This affair also defined a new strategic role for parts of Africa's vast territory and assets, a calculation of great interest to especially regionally based, ambitious Islamic Fundamentalists.[6]

Iraq's African offensive had been led on at least two occasions by the PLO's Yasser Arafat, who returned from sub-Saharan Africa embarrassed and empty handed. Once the Gulf War broke out, most of Iraq's substantial public support came from North Africa and the Sahelian belt. Overt enthusiasm was visibly expressed by the masses, though most governments in this region officially backed the United States-led alliance, investing, no doubt, in future strategic rewards. Notable in this little publicized episode is the rapid transformation of extracontinental interest in Africa that had shifted very abruptly from southern to northern and north-eastern Africa. With respect to geostrategic factors, the Red Sea is now but a mere lake in the newly expanded Middle Eastern battlefield, making at least the northern tier of Africa an integral player in the never-ending, hottest conflict zone in the world.[7]

Africa's Strategic Transition

These are very disturbing developments and are insufficiently understood, stemming no doubt from America's traditional perception of primarily Soviet-allied aggressive activities in Africa. The balancing of Soviet initiatives in Africa by Americans was a carefully managed, moderately successful process that helped keep the respective conflicts from spreading beyond a regional level. However, a major war involving Middle East-originated causes might engulf most of the northern third of the African continent, and degenerate into a confrontation between Africa's populous Arab-dominated states and black Africans in a wide belt stretching across Africa. Such an eventuality portends some tough choices for the United States which has been perplexed by developments in Libya, Algeria, Egypt and Sudan – all of which can spark sensitive and volatile changes in northern Africa. Indeed, events in northern Africa could also spill into the Middle East.

At the time that the Cold War ended and ushered in the new era, except for the Soviet Union, other interventionist states in Africa included the Cubans, whose somewhat curiously motivated activities in Africa have been abruptly terminated with no visible, lasting gains. North Koreans had maintained a non-productive, spotty presence in Africa, while China's activities had always been more constrained and narrowly focused largely on countering the Soviet Union's activities. However, India had broadened her African economic base, selling arms, and has attempted to extend her military training facilities to several African countries, mostly in East Africa, but Namibia and Nigeria as well. Africans with transnational ambitions include those who had in the past been active in their respective sub-regions, such as Libya, Morocco, Ghana, Zaire, Uganda, Rwanda, Tanzania, and South Africa. Nigeria has periodically exercised her military in the region and, most recently, played a controversial role as the overwhelming leader in the ECOMOG 'peacekeeping' force in Liberia and Sierra Leone.[8] Of interest to the United States, any future interventionist activities by Africa's own expansionist states such as Uganda, Burkina Faso, Sudan, Ethiopia, etc., will pose no major international security threat on the scale of the periodic upheavals in the Middle East's conflicts. The mid-1998 border dispute between Ethiopia and Eritrea may portend the course of such focused, armed conflicts. South Africa, even without her previous nuclear capability, remains a formidable power, strengthened by a 23-year armed conflict with SWAPO and experience gained by a prolonged semi-conventional war that was waged in tactical alliance with UNITA against Angolan government forces. But, having voluntarily yielded Namibia's independence, and turned over Walvis Bay to Namibia, there is currently no conceivable reason for the new South African government to undertake an offensive expedition in the sub-region.

Assessing Africa's current security environment as it may concern America's strategic interests is challenging due to the abrupt shift that has occurred from the traditional Cold War context to an entirely new geostrategic focus. We never did understand just what the Soviet Union's interests were in Africa, especially in Angola, that justified the outlay of such large sums. The US countered those ill-defined efforts with moderate success – despite being handicapped in assuming a mostly reactive position. But present trends concerning new interventionists especially of Middle Eastern origin, who are carving up Africa and introducing new political alignments, have the effect of greatly broadening the Middle East conflict cauldron. For example, should Iraq ever succeed in toppling the established royalist regimes on the Arabian peninsula, we may expect that immediate ramifications will quickly be felt throughout North and North-East Africa. This could spark a devastating series of conflicts which would further divide this broad region of Africa along cultural, racial, class, and religious lines – much as it is the case in Sudan. Should the eventual outcome favor those Africans inclined towards supporting Iraq's republican Ba'athist Islamic structures or the Iranian-inspired Fundamentalists, America's policies with respect to Israel and most of the Mediterranean will encounter a need for a total reassessment of Africa's new continental security challenge. Once again, we will be reacting to events in Africa; this time in response to the initiatives of Middle Eastern strategic disturbances.

America's European alliance partners also maintain their long-established African ties, usually to their former colonies. This is encouraged by the United States as it attempts to maintain an allied representation in Africa, especially in certain volatile states with scant current United States interests, but states which could become key future players.

Regional vs. Continental Considerations

The thrust of this book has been the investigation of United States relations with Africa within a Post-Cold War, regional context. We expect that a general impression was made regarding the paucity of America's historical relations with especially Central, and East Africa. North Africa determined America's distant historical security interests during World War II and the period following, which includes the protection of NATO's southern flank; the security of Israel, and to a growing extent select pro-Western, conservative Arab monarchies; and regional stabilization to ensure unimpeded access to oil – the essential ingredient that fuels the economies, strengths, and unity of the OECD developed states. Southern Africa had attracted America's greatest interest due to its respectable quantities of minerals (including some judged to be of strategic value), substantial numbers of white settlers who modernized vast parts of that region, its physical location at what had in the past been labeled a 'choke-point' for commercial and wartime navigation; and the major thrust of allied Cuban/Soviet political and geostrategic interests in the region. West Africa might have been expected to draw the most intense interest by Americans because historically, it is the source of most of the African-American community. Liberia, due to unique historical reasons, did enjoy a nebulous 'special relationship' with certain private American religious institutions and, to a lesser degree, with the United States government. And our West African interests also include the steady purchase of Nigeria's oil – a source of long-standing frustrations in dealing with that regionally prominent country's succession of military governments.

Excepting the southern Africa region with its somewhat over optimistic profile as a promising 'emerging market', due to the evidently limited other Post-Cold War American interests on the continent, there is scarce reason for the United States to approach Africa primarily on a regional basis. The Cold War ranged widely over all of Africa's regions, and this transregionalism promises to be the case in America's future incursions in Africa should the continent play an expanded supporting role for extracontinental interests. This was amply demonstrated in the contest for alliance during the war between Iraq and Iran in the 1980s. Essentially, dealing at a sub-regional level on matters such as economic integration or peacekeeping operations will be more productive for Africans than for the United States. But major external developments in which African territory may play a vital role will no doubt react as a large unit as the entire continent's strategic advantages will have to be exploited. We may deduce that until the five regions have been substantially modernized and effectively integrated into distinct, transnational units, Africa's own responses to the need for collective action will continue mainly through the weak, fragmented, and insufficiently supported continentwide organization, the Organization of African Unity.

Africa and the Clinton Administration

On assuming the presidency, the Clinton administration had been expected to redefine radically the directions and policy means of mutual United States relations with sub-Saharan Africa. With the Soviets having dropped out of Africa's diplomatic equation, Clinton's election to the presidency was timely and opportune for the Africans who had been less than enamored with the Nixon-Kissinger-Reagan-Bush Cold War axis. However, where in the past, sympathy with Africa's humanitarian plight for America's Democratic Party top office holders was generally synonymous with primarily financial largesse, Clinton's second term yielded evidence that monetary aid solved little permanently – and therefore advice concerning distinctly more conservative political and economic practices was proffered. Soon, talk concerning democratization, elections, human rights, market economies, financial discipline, and free press and speech emerged from Washington – appearing as a functional substitute for traditional aid programs. Little new was said; but it was said under the current non-Cold War circumstances in which Africa's globe-wide bargaining power had been decimated. Politically, Clinton's policy specialists identified the new suspects as Post-Cold War enemies emerging from Africa: terrorism, narcotics, crime, environmental damage, diseases, the spread of weapons of mass destruction, and the hostile, regional activities of what has come to be known as the 'rogue states', especially Libya and Iran. Sudan is identified as a state that supports terrorists. (Curiously, rarely is Iraq identified by that term with respect to that state's activities on the African continent.) Africa may still be identified (correctly) as a magnet that attracts foreign interventionists with a history of questionable interests, and as usual, with none demonstrating a positive return on their activist investments.

Altogether, the Clinton Administration's Africa posture has assumed a surprisingly low-level, conservative stance.[9] The cause for this may have been engendered by America's lending attention to the strategically more important countries of NATO's traditional bailiwick, Central and Eastern Europe; by assuming leadership of the Bosnian-Serb cauldron – which shows signs of long-term commitment; and by the undisputed realties of Africa's modest, non-competitive volume of economic relations in the newly energized global marketplace. Hence Africa's appreciation of any attention by the West, especially by the United States.

Within this context, in March 1998, President Clinton led a sizeable entourage through Africa on a 12-day trip – certainly a notable expenditure of time for an American president. The evident reasons for doing so were difficult to divine, which may suggest that political purposes, certainly domestic, may have been at the center of the effort. Notably, numerous prominent African-Americans, in and out of the Clinton Administration, participated in the trip. The determination of countries to be visited appears to have been a major challenge as care was taken not to reward countries with dubious political records. Some leaders of the countries chosen for a visit had less than impeccable credentials, but they could boast of at least some moderate recent economic or political achievements. Meetings with Heads of State with embarrassing backgrounds were conducted in private while in Uganda, as in the case of Kenya's President Daniel Arap Moi and the Democratic Republic of Congo's Laurent Kabila.[10] Ghana was chosen for Clinton's visit because that country's President Rawlings is credited with successful economic reforms in line with the punishing terms imposed by the West's financial lending institutions, and because the positive strides made for increased democratic behavior. Outside Ghana, however, most do not recall that as a military man, Rawlings had taken over the government twice by *coup d'état*, and he is currently in his fourth term in office. On assuming power in 1979, Rawlings ordered the execution of three former heads of state, as well as subsequent *coup* leaders who failed in their attempt to oust Rawlings when he was in power. His arrogant self-perception periodically manifests itself overtly and on one recent occasion included a rather less than democratic physical assault on his own 68-year old Vice-President Arkaah at a stormy Cabinet meeting.[11]

Uganda's President Yoweri Museveni, another host to the American entourage, is generally given much credit for turning that country around from the economic and political catastrophes inflicted on the country by one of Africa's most notorious presidents, Idi Amin. Like Rawlings, Museveni also assumed power as a military man and has occupied the top position for an extended period, in his case since 1986. Museveni is lauded in Washington, and the Clinton entourage identified Uganda as a maturing success story. A standard theme offered by Clinton throughout the trip was democracy, multiple political parties, and valid elections, concepts about which, in their orthodox Western forms, Museveni has often pronounced some very harsh words.[12] He held that political parties and multiparty democracy 'has brewed nothing but chaos in Africa, dividing people by tribe, religion and region'. The veracity of that claim is not contested by this analyst; what is notable about Museveni's frequent critical attacks on democracy is the continued respect he receives in the West, which (deductively), must imply that the orthodox tenets of certain democracies may be flexibly interpreted if the economy demonstrates sustained development. Unfortunately the major data usually mask the actual beneficial distribution of welfare throughout society of such economic advances.

Other stops by the Clinton entourage encountered different problems. According to media reports, Clinton had no initial plans to visit in Rwanda but was pressured to do so by some African leaders in order to draw attention to the genocide recently experienced in that country. The United States was, of course, well aware of events in that anarchic region, as non-combat American military personnel had provided extensive humanitarian assistance to Rwandan refugees in 1994. Nevertheless, a three-hour stop was scheduled in Kigali, Rwanda, where Clinton met with officials, listened to personal horror stories, and passed $2 million to a genocide survivor's fund.[13] Such a visit, however, poses major problems to the United States. We have attempted to square pragmatism with peacekeeping and humanitarian ideology by working with the new government of Rwanda – despite the power wielders behind the scene being strongly Tutsi, the faction which comprises only 15 percent of Rwanda's population.

Will not the United States once again be identified as backing a government representing a very small minority? This is an excellent example of what has been referred to as 'situational acceptance', i.e., a certain condition may be tolerated in one country – but not in another. In this regard, America's inconsistency can be seen in our policies with respect to Rwanda, Algeria, Liberia, Botswana, Nigeria, Zimbabwe, and South Africa – examples of countries with one dominant party or faction but in which the United States demonstrates different demands. Accordingly, the United States clamored for majority rule as in South Africa but we accept as a democratic product Botswana's election results, whose elite is easily determined by the domination of one ethnic group, the

Batswana who comprise 95 percent of that country's population. In this country which the United States lauds as a developmental model, the government has been accused by the weak opposition parties of curtailing the freedom to publish, and students have entered into violent confrontations with the police.[14] And the United States tolerated events in Algeria, in which the elections in 1992 were cancelled by the military after the Fundamentalists (FIS) had garnered well over a majority of the vote. In each, the delicate Cold War strategic situation may have justified such inconsistencies, but in the post-Cold War world such policies ought not to be as easily tolerated in such a peripheral foreign policy region. Also, a full accounting of neighboring Uganda's role in the Tutsi accession to power in Rwanda, and the role of transnational Tutsi involvement in Kabila's virtually unopposed march on Kinshasa has not yet emerged.[15] Clearly the Rwanda situation is still excessively fluid; under the circumstance close ties had better be shunned.[16]

There were other problems that emerged during the trip that corroborates the contention that determining the travel itinerary was difficult to establish. The official position of the United States is to encourage actively the continent's democratization. This has presented a few notable attempts – and some moderate successes, but in general, such successes must be qualified. Popularly, Africans laud the virtues of democratic political forms, but the sparsely available energy for political reform (inevitably, ambiguous) might be better spent on measurable and sustainable economic progress, especially the provision of jobs. However, before this latter qualification is realized, stability and social security must first be assured. In Rwanda Clinton confessed that he, as well as other leaders, were not fully aware of the rapid expansion of the recent crisis, and in the future, the world must act much sooner.[17] Precisely how, when, and why, was not stipulated. But the underlying assumption of attacking the ongoing global brushfires with political rhetoric can be little more than opportunistic moralizing in the absence of a newly organized globalwide, multilateral mechanism for doing so. The United States will always be asked to explain: why did we undertake such a complex (and ill prepared) humanitarian operation in Somalia, but not in Mozambique or in Liberia?

Clinton was upbeat during the trip throughout Africa, praising the emergent democracies; enthusing about the alleged 30 percent annual return on private investments in Africa – without qualification; and lauding the promising market expansion and opportunities, especially for American businessmen.[18] Collectively, he characterized these promising developments as 'Africa's Renaissance',[19] an appellation previously utilized by South Africa's heir apparent to Mandela's presidency, Thabo Mbeki.[20] Clinton also encouraged Africans to strengthen their own institutions for peacemaking, especially at the regional level.

But the ephemeral nature of Africa's political promises soon demonstrated the power of history that exacts a price from those who would prematurely depart from reality. At the regional Summit in Uganda, a promise was signed by numerous high-ranking officials to fight corruption, expand trade, place genocide leaders on trial, and 'adopt measures to avert future massacres'.[21] Among those invited and in attendance was Ethiopian Prime Minister Melse Zenawi. Less than three months later, Ethiopia's air force was actively bombing three disputed positions in Eritrea – with whom armed conflict had broken out. And, a discomforting note broke out in South Africa. There, President Nelson Mandela openly stood by his established policy of maintaining good relations with Libya, Iran, and Cuba. This is not an agreeable decision in Washington but Mandela made it clear that he will not welcome attempts to pressure his diplomatic behavior into conformity with Washington's. These three countries had stood by the ANC during the active struggle against apartheid, and proper gratitude had to be demonstrated. Finally, analysts had questioned Clinton's choice of Botswana for a visit. Speculation for this choice turned out to be the least ambiguous of the entire trip: it was personal safari time in Chobe Game Park.

This longest trip to Africa by a United States president in modern times appears to have been composed lightly and very situationally, but to produce high visibility. Critics argued that his delicate domestic political situation would benefit from some distant relief. Still others noted an evident need to lead a primarily African-American entourage in order to develop their mutual commercial ties with Africa. Recently a shift appears to be emerging among established African-American groups to a more conservative political and free enterprise profile. Washington can effectively encourage the exploration of Africa's commercial opportunities by African-Americans with new legislative programs presently underway. Clinton spoke of diverse joint projects with Africans, power generation in Ghana, $120 million for education and Internet links in Africa, an increase in monetary aid from its present modest level, American investors are to get generous terms for loan guarantees, an aviation and an energy agreement was signed with Senegal, and the passage of a bill in Congress is to create a special free-trade zone in order to encourage and abet trade with Africa on generous terms. Still other measures would be introduced openly to facilitate the African-American business community's approach to the 'new Africa'.[22]

In the security realm, Clinton sought to further the future continent-wide reception and capability of the African Crisis Response Initiative (ACRI). The goal of this American-initiated effort is to structure 'partnerships with both regional countries and allies and friends outside the region to train fully interoperable, highly effective, rapidly-deployable African peacekeeping units capable of operating jointly'.[23] United States plans call for the strength of the Organization of African Unity to be enhanced and that subsequently, regionally based, multinational peacekeeping contingents would be available for early conflict intervention. In form, ECOMOG's complex and lengthy involvement in the attempt at pacifying Liberia has served as a precedent for other such efforts. However, due to its prolonged nature and numerous controversial operations, the final verdict on ECOMOG's utility has not yet been passed. It may well be that ECOMOG's effort will be judged an outright failure with respect to its central objective.[24]

President Clinton's approach to Africa offers no radical innovations nor policies, certainly not on the scale of the changes in America's approach to the members of the defunct Warsaw Pact. The reluctance to increase massively the aid budget derives from a long-standing perception that aid produces little, even if absorbed honestly by the receiver. Aid is viewed as building dependency; is improperly invested when channeled primarily to immediate social needs; it may encourage fiscal irresponsibility; the direct beneficiaries of aid are not necessarily those in greatest need; and the donor easily becomes fatigued on seeing no commensurate results.

A visible action that Clinton has taken concerns the large number of appointments of African-Americans to prominent positions in his administration, and his evident involvement of African-Americans on matters concerning United States–Africa relations – whether official or private interests are to be pursued. At the official level, America's Africa policy has increasingly been reviewed by African-American interests. The reasons for this are political, stimulated initially by the high profile success of prominent African-Americans involved in the mid-1980s boycott and sanctions campaigns against South Africa. The efforts of TransAfrica, an effective African-American lobby group focusing on black American interests, sensitized African-Americans to the potential of organized, extraparliamentary style, opposition activities regarding the formulation and execution of United States foreign policy.

Clinton's enthusiastic reference to the great earning opportunities in Africa and that continent's emerging 'Renaissance', can easily be challenged as excessive baroqueish hyperbole, if for no other reason than the tendency that every time such pronouncements are lavished, another calamity will soon follow. In this case, less than three months after his Africa trip, despite agreements signed at the Summit, Ethiopia bombed three border positions in a violent outbreak of hostilities with Eritrea, and shortly thereafter, Guinea-Bissau came under violent siege by its military forces. Africa is much too complex to allow for sweeping observations by occasional foreign visitors, nor will the fortunes of that continent's inhabitants be greatly affected by America's well-meaning and well-worn political and economic ideological mantras. Africa can, and no doubt will, settle down, but it will take a massive cooperative effort, as well as a new, creative, and unprecedented global commitment, for which the United States and other developed states have not yet demonstrated their readiness. It will take years to deconstruct the legacy, cost, and architecture of the Cold War that did indeed pose credible threats to the West's core security interests. By contrast, today Africa poses mostly humanitarian challenges.

Notes

1 Quoted in *Montgomery Advertiser* (9 September 1995).
2 Monte Palmer (1985) presents a useful analytic framework which distinguishes between traditional, transitional, and modern societies. *Dilemmas of Political Development* (Itasca, IL: F. E. Peacock, 3rd edn), 23 and Ch. 2.
3 That Africa's identity is determined by Europeans is the subject of a provocative chapter by Ali M. Mazrui (1986) *The Africans: A Triple Heritage* (Boston: Little, Brown), Ch. 5.
4 The theory and several African cases of such 'Prolonged Wars' are presented in Karl P. Magyar and Constantine P. Danopoulos (eds) (1994) *Prolonged Wars: A Post-Nuclear Challenge* (Maxwell AFB, AL: Air University Press).
5 The theme concerning the dire consequences of the West's ignorance of non-Western conflict cultures is provocatively presented by Adda B. Bozeman (spring 1976) 'War and the Clash of Ideas', in *Orbis: A Journal of World Affairs*, 20(1).
6 Much of the viewpoint of the emerging relationship between Islamic Fundamentalism and the 'nervous' West was elaborated by Sudan's Dr Hasan Turabi and a group of advanced academic specialists including Dr Karl P. Magyar in *Islam, Democracy, the State and the West: A Roundtable with Dr Hasan Turabi*, Arthur L. Lowrie (ed.) (1992) (Tampa, FL: The World and Islam Studies Enterprise).
7 The perceptual unity of Africa and a large part of the Middle East is best developed by Ali Mazrui (1986), 28–30.
8 See Terry Mays, 'Nigerian Foreign Policy and Its Participation in ECOMOG', in Karl P. Magyar and Earl Conteh-Morgan (eds) (1998) *Peacekeeping in Africa: ECOMOG in Liberia* (London: Macmillan), Ch. 6.
9 The White House (May 1977) *A National Security Strategy for a New Century* (Washington, DC), 27–8.

10 Reported in *USA Today* (26 March 1998).
11 Antony Goldman (10 January 1997) *The Christian Science Monitor*. Also, Nate Bowditch (6 December 1996) The *Christian Science Monitor*. Baffour Ankomah (February 1996) offers more detail regarding the assault, in 'Has Rawlings Thrown His Last Punch?,' *New African*. A tough analysis of just what is going on in Ghana and its alleged economic success is credibly challenged by Marguerite Michaels (1992/93) 'Retreat from Africa', in *Foreign Affairs*, 72(1), 99–100. See also the critical commentary on Uganda and Ghana by Alfred Oryeda (16–29 March 1998) 'From Revolutionaries to Democrats', *West Africa*, 356.
12 Reported by James C. McKinley, Jr (31 March 1996) *The New York Times*. Also, Paris AFP in English (3 July 1995) 'Assembly Extends Museveni's Rule 5 Years', Foreign Broadcast Information Service, *Daily Report*, FBIS-AFR-95-127, 5–6.
13 Reported by Mimi Hall (1 March 1998) *USA Today*.
14 Reported in *New African* (October 1995).
15 See David Aronson (8 July 1997) 'Who really controls the new government in Kinshasa?' *Christian Science Monitor*.
16 François Misser and Alan Rake (February 1997) 'The Volcanoes are Exploding', *New African*.
17 Mimi Hall (26 March 1998) 'Clinton Says World Took Too Long to Aid Rwanda,' *USA Today*.
18 Carol S. Castiel (6–26 April 1998) reported on the great effort that was made regarding the commercial nature of the entire trip but was not reported as such by the United States media, which reported much more on the proposed developmental assistance projects. This in turn kept the media spotlight away from the private sector which had initially been the intended focal point. 'Entrepreneurial Africa', *West Africa*, 391.
19 Baffour Ankomah (May 1998) 'Clinton's African odyssey', *New African*.
20 James A. Joseph (15 December 1997) 'The Idea of African Renaissance', delivered in Washington, DC, reproduced in *Vital Speeches of the Day*, 133.
21 Mimi Hall (26 March 1998) *USA Today*.
22 Karen Breslau (6 April 1998) 'African Dreams', *Newsweek*.
23 William S. Cohen (1998) *Annual Report to the President and the Congress* (Washington, DC: US Government Printing Office), 14.
24 A definitive judgment on the ECOMOG operation in Liberia should not be expected soon – yet would be a welcome contribution. The problem concerns the criterion for judgment. See, Karl P. Magyar, 'Conclusion: Liberia's Peacekeeping Lessons for Africa'.

Index